State of Washington
ARTHUR B. LANGLIE, Governor

Department of Conservation and Development
W. A. GALBRAITH, Director

DIVISION OF MINES AND GEOLOGY
SHELDON L. GLOVER, Supervisor

Bulletin No. 42

GOLD IN WASHINGTON

By

MARSHALL T. HUNTTING

OLYMPIA
STATE PRINTING PLANT
1955

For sale by Department of Conservation and Development,
Olympia, Washington. Price, one dollar.

FOREWORD

Gold, throughout the ages, has been synonymous with wealth. For thousands of years, it has been the foremost medium of exchange in most countries. Everyone is familiar with gold and with the romance associated with its recovery from occurrences in nature. In fact, it is probable that gold is the first, and to many the only, metal thought of when mining is mentioned. Certainly, it is the metal first thought of under ordinary conditions when a person decides to become a prospector and seek his fortune on the streams and in the mountains. This is easily understood, for, of all metals, gold is the most simply and easily recovered from its containing formation, whether it occurs as a lode or as a placer deposit. A minimum of experience and equipment is required, and, when won, gold is tangible wealth, requiring little or no treatment to be exchangeable for goods.

The interest in gold shown by the larger mining concerns has declined in late years because its fixed price does not compensate for the continually increasing costs of mining—in labor, equipment, and supplies. This situation will change when and if the price of gold is increased. In the meantime, small-scale gold-mining operations, with limited investment, still find gold mining attractive. That such operations are still more attractive and profitable in periods of depression is obvious, and the fear of "hard times" is the motivation for much of the interest constantly shown in gold prospecting.

The present report is designed to answer the questions most commonly asked as to how and where gold occurs in Washington, and to provide useful information to those who wish to examine old properties or search for new occurrences. It is believed that the report will prove useful and repay the time and effort that has been necessary in its preparation.

SHELDON L. GLOVER, Supervisor.
Division of Mines and Geology

February 17, 1955

CONTENTS

ILLUSTRATIONS

GOLD IN WASHINGTON

By Marshall T. Huntting

INTRODUCTION

Gold has a wider appeal to prospectors than almost any other metal, and although other metals, uranium for example, have attracted much popular attention at times, it is likely that gold will continue for many years as a staple in the field of prospecting. In the inquiries received by the Division of Mines and Geology, gold has been the metal most frequently asked about, and the purpose of this report is to provide some of the data most often requested, including a list of the more important gold occurrences, both lode and placer, in the state of Washington. Most of the general data in this report are from various published sources, which are indicated in the list of references. The section on occurrences is abstracted from the chapters on lode and placer gold in Bulletin 37, Inventory of Washington Minerals, Part 2, Metallic Minerals, a report that is in preparation in 1955 by the Division of Mines and Geology, but the publication date of which is yet undetermined.

SELECTED BIBLIOGRAPHY

Arnold, Ralph, Gold placers of the coast of Washington: U. S. Geol. Survey Bull. 260, pp. 154-157, 1905.

Averill, C. V., Placer mining for gold in California: California Div. Mines Bull. 135, 1946.

Bartlett, Clare, The Pacific Coast beach sands: Mines and Minerals, vol. 30, p. 375, 1910.

Baxter, C. H., and Parks, R. D., Mine examination and valuation, 2d ed., pp. 18-77, Houghton, Michigan College of Mines and Technology, 1939.

Blackburn, C. F., Gold in the Cascades: Mining and Scientific Press, vol. 61, p. 91, and vol. 60, p. 412.

Boericke, W. F., Operating small gold placers, 2d ed., John Wiley & Sons, Inc., 1936.

Browne, J. R., Resources of the Pacific slope, New York, D. Appleton and Co., 1869.

Butler, G. M., Some facts about ore deposits: Arizona Bur. Mines, Geol. Ser. no. 8, Bull. 139, 1935.

Collier, A. J., Gold-bearing river sands of northeastern Washington: U. S. Geol. Survey Bull. 315, pp. 56-70, 1907.

Cooper, C. L., Mining and milling methods and costs at Knob Hill mine, Republic, Washington: U. S. Bur. Mines Inf. Circ. 7123, 1940.

Culver, H. E., Geology of Washington—Part I, General features of Washington geology (with preliminary geologic map in colors): Washington Div. Geology Bull. 32, 1936.

Day, D. T., and Richards, R. H., Investigations of black sands from placer mines: U. S. Geol. Survey Bull. 285, p. 156, 1906.

..............................., Useful minerals in the black sands of the Pacific slope: U. S. Geol. Survey Min. Resources 1905, pp. 1175-1258, 1906.

Gardner, E. D., and Johnson, C. H., Part I, General information, hand shoveling, and ground sluicing: U. S. Bur. Mines Inf. Circ. 6786, 1934.

........................, Part II, Hydraulicking, treatment of placer concentrates, and marketing of gold: U. S. Bur. Mines Inf. Circ. 6787, 1934.

........................, Part III, Dredging and other forms of mechanical handling of gravel, and drift mining: U. S. Bur. Mines Inf. Circ. 6788, 1935.

Gardner, E. D., and Allsman, P. T., Power-shovel and dragline placer mining: U. S. Bur. Mines Inf. Circ. 7013, 1938.

Ingersoll, G. E., Hand methods of placer mining and placer mining districts of Washington and Oregon: State College of Washington Eng. Exper. Sta. Bull. 40, 1932.

........................, Small scale methods of placer mining and placer mining districts of Washington and Oregon: State College of Washington Eng. Exper. Sta. Bull. 43, 1933.

Jackson, C. F., and Hedges, J. H., Metal-mining practice: U. S. Bur. Mines Bull. 419, 1939.

Jackson, C. F., and Knabel, J. B., Sampling and estimation of ore deposits: U. S. Bur. Mines Bull. 356, 1932.

Johnson, F. W., and Jackson, C. F., Federal placer-mining laws and regulations and small-scale placer-mining methods: U. S. Bur. Mines Inf. Circ. 6611R, 1938.

Kelly, J. V., Columbia River magnetite sands, Clatsop County, Oregon, and Pacific County, Washington. Hammond and McGowan deposits: U. S. Bur. Mines Rept. Inv. 4011, 1947.

McKinstry, H. E., Mining geology, New York, Prentice Hall, Inc., 1948.

Masson, D. L., Small scale placer mining: Washington State Inst. Tech. Bull. 45 M, 1953.

Mertie, J. B., Jr., Placer gold in Alaska: Jour. Washington Acad. Sci., vol. 30, no. 3, pp. 93-124, 1940.

Pardee, J. T., Platinum and black sand in Washington: U. S. Geol. Survey Bull. 805, pp. 1-15, 1929.

Peele, Robert, Mining engineers' handbook, 3d ed., New York, John Wiley & Sons, Inc., 1941.

Rickard, T. A., Sampling and estimation of ore in a mine, New York, Hill Publishing Co., 1907.

Sharwood, W. J., and von Bernewitz, M. W., Bibliography of literature on sampling: U. S. Bur. Mines Rept. Inv. 2336, 1922.

Soderberg, R. L., Mining methods of the Holden mine, Howe Sound Co., Chelan Division, Holden, Washington: U. S. Bur. Mines Inf. Circ. 7448, 1948.

Stevens, I. I., Explorations for a railroad route from the Mississippi River to the Pacific Ocean: Pacific Railway Reports, vol. 12, no. 1, 1860.

Trimble, W. J., The mining advance into the Inland Empire: Univ. Wisconsin Bull. 638, History Ser., vol. 3, no. 2, 1914.

Van Nuys, M. H., An outline of mining laws of the State of Washington: Washington Div. Mines and Geology Bull. 41, 1953.

Von Bernewitz, M. W., Handbook for prospectors, 4th ed., New York, McGraw-Hill Book Co., Inc., 1943.

PROPERTIES OF GOLD

Gold is bright yellow when pure, but, as originally found, most gold is naturally alloyed with silver, and the color varies through lighter shades of yellow to yellowish-white as the silver content increases. Much less commonly gold is found naturally alloyed with copper. Such alloys are reddish yellow. Gold has also been found naturally alloyed with iron, platinum, bismuth, and mercury. Although gold does not tarnish, nuggets and flakes of the metal are often covered by reddish, brown, or black films of iron and manganese oxides.

Gold is the most malleable and ductile of metals and is one of the softest. Its hardness is 2.5 to 3—harder than lead but softer than copper or silver. It can be hammered out into leaves thin enough to be transparent and so thin that more than 500 of them would be required to equal the thickness of this page; also it can be drawn out in wire so fine that one troy ounce would produce a wire more than 57 miles long. Gold weighs 1,206 pounds per cubic foot, and the specific gravity of pure gold is 19.32 at 17.5° C. However, the specific gravity of natural gold may be as low as 12.5, owing to the presence of alloy metals. Gold is a good conductor of heat and electricity, its electrical conductivity being exceeded only by silver and copper. Its electrical resistivity is 2.4×10^{-6} ohms per cubic centimeter. The melting point of the metal is 1063° C., and the boiling point, 2600° C. The thermal coefficient of linear expansion of gold at 20° C. is 14.2×10^{-6}, and the magnetic susceptibility at 18° C. is -0.15×10^{-6}.

Gold crystallizes in the isometric system, but distinct crystals are comparatively rare. Usually those found are small, more or less distorted, and in dendritic or branching groups. The most common crystal forms are cubes, octahedrons, and dodecahedrons, but usually gold occurs as disseminated scales or grains, as filiform, dendritic, spongy, or reticulate particles, or in larger irregular lumps or nuggets. The individual particles are commonly submicroscopic in size but may range up to nuggets weighing several pounds each. Crystalline gold has been recovered in Washington from some of the lode deposits in the Swauk district of Kittitas County.

The chemical symbol for gold is Au, the atomic weight is 197.2, and the atomic number is 79. Gold is both univalent and trivalent, but is not very active chemically and forms very few compounds. It is one of the most permanent and least active of the metals; because of its resistance to oxidation the alchemists called gold a "noble metal" in contrast to the base metals, which oxidize when heated in air. Gold is not affected by air or most reagents and does not dissolve in any of the common single acids but does dissolve in aqua regia, a combination of hydrochloric and nitric acids. Also, it is soluble in alkali cynides in the presence of air.

USES

Gold has been desired and used by man from before the beginning of written history. Its color, ease of recovery from its ores, extreme malleability, chemical stability, and rarity have combined to create man's early and continued interest in the metal, and these same properties account for its preeminence in its current uses for monetary and ornamental purposes. Since earliest historical time gold has been used for currency or as a monetary standard, and at the present time these are the principal uses of the metal. Gold holdings of the United States Treasury on December 31, 1950 were $22,706,000,000. This is a decrease of 7 percent from the previous year and represents a reversal of the trend of increasing gold reserves which had been in effect since 1946.[1]

In the arts gold is used in the manufacture of jewelry, watches, and gold foil for lettering and decorative purposes. Lesser amounts are used in dentistry and in the electrical and chemical industries. Because of their nonoxidizing characteristics at high temperatures, gold alloys are used in some special electrical contact points; also, on account of its high resistance to chemical attack, gold is used in technological and laboratory equipment used for handling corrosive fluids. Gold-platinum alloys are used in the textile industry in spinnerets which form the filaments of rayon and other synthetic fibers. Small quantities of gold are used in medicine and photography.

For most of its uses gold is alloyed with other metals, generally to increase its hardness. The metals commonly used in the alloys are copper, silver, nickel, tin, aluminum, iron, zinc, lead, platinum, and palladium. Gold-mercury alloys are known as amalgams. The purity of gold is expressed as "fineness," based on 1,000. For example, gold that is 915 fine is 91.5 percent pure. However, jewelers generally express the gold content of an alloy in "carats." Pure gold (1,000 fine) is called 24 carat. Gold that is marked 20 kt. is 20/24 pure gold, or 833 fine, or 83.3 percent pure gold by weight. The unit of weight used for gold is the troy ounce, which is equal to 1.097 avoirdupois ounces, 31.103 grams, or 480 grains.

GOLD ORES AND ORE MINERALS

Although gold is one of the scarcer elements, it is widely distributed in nature and has been found in minute quantities in various kinds of igneous, metamorphic, and sedimentary rocks in places remote from known ore deposits. Gold is known to concentrate in certain plants and in some marine animals. The ashes of *Equisetum* (commonly called the "Horsetail" rush) have been reported to contain up to 610 grams of gold per ton.[2] Sea water has been variously reported to contain from 5 to 65 milligrams of gold per ton,[3] and on the basis of such assays one estimate of the total content of gold in all the oceans is 10,000,000,000 tons.

Gold occurs by far most commonly as the native metal, which always is alloyed with varying amounts of silver, generally 10 to 20 percent. Any alloy in which the silver content exceeds 20 percent is known as electrum. Usually the gold in ores is in particles too small to be seen with the unaided eye; commonly it is of submicroscopic size. Other than the native metal and its alloys, the only naturally occurring gold minerals are the tellurides, the most common of which are calaverite, $AuTe_2$, containing 43.5 percent gold; sylvanite, $(Au,Ag)Te_2$, containing 24.5 percent gold; and petzite, $(Au,Ag)Te$, containing 25.2 percent gold. These minerals are rather rare, and none of them has been recognized in Washington. Other gold minerals even less common are the gold-silver-mercury tellurides, kalgoorlite and coolgardite; the silver-gold tellurides, muthmannite and goldschmidtite; the gold telluride, krennerite; and the gold-lead-sulpho-telluride, nagyagite.

Gold accompanies selenium and particularly tellurium, the latter association being illustrated by the gold tellurides in sulfide ore deposits. In the presence of both sulfides and arsenides or antimonides gold usually is found concentrated in the arsenides or antimonides rather than in the sulfides.[4] In sulfide deposits gold is usually associated with pyrite and less commonly with arsenopyrite; it may be found also in chalcopyrite, stibnite, pyrrhotite, sphalerite, and other sulfides. The most common gangue mineral is quartz, but other gangue minerals in gold ores include carbonates, fluorspar, tourmaline, and barite.

Gold deposits are of three main types: veins or other lode ore bodies of hydrothermal origin, ordinary placer deposits, and consolidated placer deposits (gold-bearing conglomerates and sandstones).

---

[2]Rankama, Kalervo, and Sahama, Th. G., Geochemistry, University of Chicago Press, p. 707, 1950.

[3]Clarke, F. W., The data of geochemistry: U. S. Geol. Survey Bull. 770, pp. 124-125, 1924.

[4]Rankama, Kalervo, and Sahama, Th. G., op. cit., p. 705.

LODE DEPOSITS

In hydrothermal deposits gold occurs in recoverable quantities in most ores of silver, copper, bismuth, and antimony and in many ores of lead and zinc, as well as in many deposits where gold is the only value. These deposits range from deep-seated, high-temperature ores to epithermal ores, and they may be found in rocks of many types, but most commonly gold-bearing ores are in quartz veins closely associated with granitic or volcanic rocks of acidic to intermediate composition. Because gold is chemically inert it is sometimes found to be concentrated as free-milling ore in the oxidized zones of sulfide bodies through leaching and partial removal of the sulfides. Many mines have operated successfully on such free-milling ore, but as the workings went to greater depths and reached the base (sulfide) ore the operations became unprofitable, largely because of increased cost for metallurgical treatment of the base ore. Lode gold ore mined in Washington in 1950 averaged 0.485 ounce of gold and 1.509 ounces of silver per ton, as compared with averages of 0.277 ounce of gold and 0.273 ounce of silver per ton for the United States as a whole.

PLACER DEPOSITS

Because gold has a high specific gravity and is chemically inert it becomes concentrated in placer deposits, where it commonly is found associated with magnetite, ilmenite, chromite, monazite, rutile, zircon, garnet, and other heavy minerals. These minerals are the principal constituents of the so-called "black sands." The gold-bearing sands and gravels in placers may be derived from lode gold deposits, but the gold in many placers originated not in lode gold ores but as sparsely disseminated gold in rock too lightly mineralized to be classed as ore. Because of this and the fact that lode deposits from which some placers have been derived have been completely destroyed by erosion, a search for the "mother lode"— the "hard-rock" source of the gold—may be futile in many placer districts. Placer deposits may be of many kinds, as (1) residual deposits from weathering of rocks in place, (2) river gravels in active streams, (3) river gravels in abandoned and often buried channels, (4) eolian deposits, (5) ocean beaches at sea level, and (6) ancient ocean beaches now raised and inland.

Beach placers form through the agency of ocean waves reworking beach sands and gravels and concentrating the heavy minerals. These sands and gravels may represent alluvium brought to the coast by streams, or they may originate from the erosion of the bedrock of the sea cliffs or erosion of unconsolidated glacial or other sediments which overlie the bedrock along the shore. Gold present in minute amounts in the eroded material is released by this process and accumulates with other heavy minerals in the beach sands.

Beach placers are frequently found to be enriched when examined directly following storms, when wave action has been especially vigorous.

Stream placers usually are restricted to present-day valleys, but there are numerous examples of bench placers in bars and terraces high above present stream channels and of placers in old valleys not now occupied by streams. Gold in streams usually accumulates at places of slackened stream velocity, as in a broader valley below a narrow gorge, at junctions of tributaries, near the heads of quiet reaches, and on the inside of bends. Furthermore, placer values are almost always concentrated on or near bedrock or on a "false bedrock," which may be a hard clay bed.

Consolidated placers are formed through the same processes that produce ordinary unconsolidated placer gravels and sands, but the consolidated placer deposits originated in pre-Pleistocene time and have subsequently been converted to conglomerates, sandstones, and even quartzites by cementation, compaction, and sometimes recrystallization. The beds comprising these deposits usually are overlain by other sedimentary rocks or by flows of volcanic rocks.

PROSPECTING FOR GOLD

As the old saying goes, "gold is where you find it," and the places where it is found are many and varied. Some very unlikely looking rocks carry gold in paying quantities, and other rocks that are mineralized and look very favorable for the occurrence of gold have none. However, there is another old saying that advises the prospector to "look for bears in bear country." In other words, the best areas in which to prospect for gold are those where gold has been found previously. This is axiomatic, of course, yet it is surprising how much effort is wasted each year by people who, for reasons of their own, persist in prospecting some of the least likely areas. The maps (figs. 1 and 2 facing pp. 111 and 112) show the locations where lode and placer gold have been found in the state, and, although in the future a few discoveries will no doubt be made in areas remote from previously known occurrences, the great majority of the new finds will be in areas shown on the maps where gold has been found most commonly in the past. Thus, the maps are probably the best, and yet simplest, guides to prospecting this report can furnish.

As noted before, there are few rock types that may be eliminated as possible hosts for gold ore deposits, but most lode gold is in quartz veins closely associated with intrusive or extrusive igneous rocks of intermediate to acidic composition, such as diorite, granodiorite, granite, andesite, or rhyolite. This fact suggests, as an aid in prospecting for gold, the use of geologic maps showing the areas underlain by these rocks. A geologic map covering the whole state[5] has been published, as has an index map[6] which shows the outlines of all the geologically mapped areas in the state and cites the literature in which these maps may be found.

The application of geological principles and the study of geologic maps and aerial photographs may be of assistance in both placer and lode prospecting. An insight into the probable location of channels and paystreaks in placers may be gained through a study of physiography and the application of stream-sedimentation principles. As placer gold tends to concentrate in stream channels in places of slackened water velocity, the prospector should examine with special care such places as the bars on the inner sides of curves and places where streams emerge from rapids into quiet water. Attention to the structure and the types of bedrock may be profitable, as gold often concentrates on the edges of steeply dipping schist, thin-bedded sediments, and other formations that erode into miniature ridges and grooves, thus forming natural riffles in the stream bed. Riffles parallel to the direction of streamflow are more effective than riffles crossing at a large angle;[7] therefore, portions of channels roughly paralleling the strike of such rocks are most favorable. The pitted, rough surfaces of limestone are more effective in catching gold particles than are the smooth surfaces that erosion produces on most granitic rock, although under favorable conditions potholes may form in granite as well as other rock types, and these potholes may act as effective gold traps.

Oftentimes attention to details in placers may lead the prospector to the discovery of lode deposits. Gold particles in stream gravels commonly may be traced upstream to their source even though the placer itself may be too low in grade to work at a profit. The size and shape of the gold particles give clues to the distance they have traveled. Nuggets found near their point of origin may be comparatively large and quite irregular and angular in shape, whereas particles that have been transported for some distance generally are smaller, more regular, and rounded or flattened.

Geophysical investigations, using magnetic, electric, gravimetric, or seismic methods, are sometimes helpful in locating mineralized zones in bedrock or, when placer prospecting, in determining depth to bedrock or in locating magnetic black sands with which gold may be associated. Geophysical exploration methods and techniques are described in detail by Heiland[8] and Jakosky.[9] The prospector

[5]Culver, H. E., Geology of Washington—Part I, General features of Washington geology (with preliminary geologic map in colors): Washington Div. Geology Bull. 32, 1936.

[6]Boardman, Leona, Geologic map index of Washington: U. S. Geol. Survey map, scale = 1:750,000, 1949.

[7]McKinstry, H. E., Mining geology, New York, Prentice Hall, Inc., p. 227, 1948.

[8]Heiland, C. A., Geophysical exploration, New York, Prentice-Hall, Inc., 1940.

[9]Jakosky, J. J., Exploration geophysics, Los Angeles, Times-Mirror Press, 1940.

should be very careful to distinguish between geophysical prospecting equipment (all of which has been described in published technical papers) and the many and varied kinds of "doodlebugs," which operate on "mysterious" and "secret" principles known only to the operator or to no one. These include many variations on the old-time pendulum-type doodlebug, the forked divining rod or witch stick, and the "electronic" and other more imaginative and modern versions, all of which have been represented by misguided or unscrupulous persons to be infallible in locating, identifying, and evaluating ores. These same gadgets are usually reputed to be equally effective in locating water, oil, buried treasure, lost persons, and just about anything imaginable. This "divining rod myth" is effectively disposed of by Butler,[10] who also discusses a number of useful facts about ore deposits in a paper written especially for prospectors and miners.

SAMPLING

In searching for lode-gold ores the prospector is obliged to depend heavily upon assays to determine whether or not a vein carries gold values or how rich a given ore is. Until a person becomes very familiar with the gold ore in a given mining district his estimate of the value of a specific ore sample can be little more than a guess, and in many gold mines even experienced operators have to rely entirely upon assays to distinguish between ore and waste. Upon finding a possible gold-bearing vein the prospector may be well advised to "high-grade" the deposit and take his preliminary samples from the best looking portions of the ore. Then if the assays show nothing or only traces of gold the deposit may be eliminated from further consideration. If the assays are favorable the vein should be resampled in such a way that the samples will be as truly representative as possible. Details of sampling methods and of evaluating ore deposits, both lode and placer, are described in many books,[11] and a bibliography of literature on sampling has been published by the U. S. Bureau of Mines.[12] Sampling methods, as well as most of the other subjects that would be of interest to the gold prospector, are also described in a very usable handbook for prospectors by Von Bernewitz.[13]

[10]Butler, G. M., Some facts about ore deposits: Arizona Bur. Mines, Geol. Ser. no. 8, Bull. 139, pp. 71-77, 1935.
[11]Baxter, C. H., and Parks, R. D., Mine examination and valuation. 2d ed., pp. 18-77, Houghton, Michigan College of Mines and Technology, 1939.
Jackson, C. F., and Knabel, J. B., Sampling and estimation of ore deposits: U. S. Bur. Mines Bull. 356, 1932.
Rickard, T. A., Sampling and estimation of ore in a mine, New York, Hill Publishing Co., 1907.
Peele, Robert, Mining engineers' handbook, 3d ed., New York, John Wiley & Sons, Inc., 1941.
McKinstry, H. E., Mining geology, New York, Prentice Hall, Inc., 1948.
[12]Sharwood, W. J., and von Bernewitz, M. W., Bibliography of literature on sampling: U. S. Bur. Mines Rept. Inv. 2336, 1922.
[13]Von Bernewitz, M. W., Handbook for prospectors, 4th ed., New York, McGraw-Hill Book Co., Inc., 1943.

Placer sampling is usually accomplished by panning, but rockers or even small sluice boxes may be used. Since placer gold tends to concentrate on or near bedrock, it is necessary in most places for the prospector to dig trenches, pits, or shafts, or to drill to obtain samples for testing. The following description of the technique of panning may be of some assistance to the beginner.

The pan of gravel is held under water, the contents are stirred by hand to break up cemented pieces of gravel and lumps of clay, and the larger stones are picked out. The pan, still under water, is held with both hands in a level position and is given a series of rotary or gyratory motions, largely through wrist action. This causes the heavy particles to settle and the light material to work its way to the top. The pan is tilted, and by raising and lowering it in the water the surface material is washed off. Then by holding the pan in a position tilted away from the prospector, so the contents just barely do not spill, it is alternately gyrated to bring the light material to the surface and washed by raising and lowering the tipped pan in the water. As they come to the surface the larger pieces of barren gravel are picked off by hand. The panning is continued with gradually increased tipping until only the gold and a little heavy sand are left. The residual sand may be separated from the gold by drying and blowing, by removal of magnetite with a magnet, or by adding mercury to collect the gold as amalgam. In experienced hands there is little or no loss of gold.

SIZE OF GOLD PARTICLES

The size of individual gold particles ranges widely. Commonly the gold in lode deposits occurs in particles of submicroscopic size. In placer deposits the finest flour gold may be too tiny for individual particles to be seen without the aid of a magnifier, but they may range upward in size to nuggets weighing several pounds each. The largest nugget known is the Welcome Stranger, weighing 190 pounds, found in 1869 near Ballarat in Australia. The largest nuggets yet found in Washington are from the Swauk district in Kittitas County. Landes[14] reports a $1,100 nugget that was found in 1900 at the Elliott placer on Williams Creek, and a $1,004 nugget on a bench of Swauk Creek above the mouth of Baker Creek.

Placer gold particles are classified as (1) coarse—more than 0.06 inch in diameter (about the size of a grain of rice), (2) medium—less than 0.06 inch but more than 0.03 inch in diameter (about half the size of a pinhead), (3) fine—less than 0.03 inch but more than 0.015 inch in diameter (about a quarter of the size of a pinhead), and (4) very fine—less than 0.015 inch in diameter. Fine gold averages 12,000 colors (particles) per ounce, and very fine averages 40,000

---

[14]Landes, Henry, Thyng, W. S., Lyon, D. A., and Roberts, Milnor, The metalliferous resources of Washington, except iron: Washington Geol. Survey Ann. Rept. for 1901, pt. 2, p. 88, 1902.

colors per ounce. Gold particles that require 300,000 or more colors per ounce are called flour gold, and about 100 particles of this size are required to have a value of one cent. Some flour gold is so small that it takes 4,000 colors to be worth one cent, and about 14 million colors are needed to weigh one ounce. Tiny as they are, each of these individual particles can be seen when placed on a black surface. A particle of gold the size of a common pinhead has a value of about one cent; a particle the size of a grain of rice has a value of about 20 cents; and a particle the size of a navy bean has a value of about $3.50. One troy ounce of pure gold (worth $35) has a volume equal to that of a cube a little less than half an inch (0.464 in.) on a side.

IDENTIFICATION TESTS

Many times the novice prospector is undecided whether the "yellow stuff" he is looking at is really gold or is something else. The yellow minerals that are most commonly mistaken for gold are pyrite, chalcopyrite, and golden-colored mica flakes. Pyrite, or "fool's gold," is heavy, but not as heavy as gold; it is hard and brittle and crushes to a black powder when hammered, whereas gold is soft (almost as soft as lead) and malleable and can be easily beaten into very thin sheets that are flexible (can be bent a number of times without breaking). Pyrite is soluble in concentrated nitric acid; gold is insoluble. Chalcopyrite, also sometimes mistaken for gold, is similar to pyrite in these properties. Pyrite commonly occurs as cubic crystals, but gold almost always is found in irregular shapes, and in those rare places where it does occur as crystals the crystals are always in intergrown masses.

Tiny golden-colored mica flakes sometimes look deceptively like gold, but the luster of mica is different from that of gold; mica has laminations that can be split with a knife; and mica flakes, like gold, are flexible, but, unlike gold, the flakes are elastic, so that when bent they tend to return to their original shape. Gold is malleable, but mica is not; when mica is hammered it breaks up into numerous tiny flakes. Gold is heavy, but mica is light. Thus, when panned, gold becomes concentrated in the very lowest part of the pan, but mica will be washed out of the pan, although, because of its flakiness, it does tend to segregate somewhat from other light minerals. Mica is difficultly fusible; gold, pyrite, and chalcopyrite fuse easily in a blowpipe flame (gold at 1063° C.); and gold when roasted is odorless, but the sulfides, pyrite and chalcopyrite, yield sharp-smelling sulfur dioxide fumes.

The gold telluride minerals are comparatively rare and are not easily recognizable. Furthermore, they usually occur as small, sparsely disseminated grains that are difficult to isolate for testing. They vary in color from silver white, yellow, and steel gray to nearly black. Gold in the tellurides may be recognized by its physical properties after the tellurides have been roasted, but it can best be detected by fire assay.

There are no simple, easily performed chemical tests for gold, but Scott[15] describes several qualitative tests as well as several wet assay methods. The "Purple of Cassius" test, often referred to as the most easily performed test for gold, may be satisfactory as a means of confirming the identification of an unknown metal suspected of being gold, but as a means of detecting gold in an ore it can not ordinarily be relied on.

Spectrographic methods are not very satisfactory for the detection of gold because the gold concentration in many ores is below the limit of spectrographic observation. Ahrens[16] discusses the problems of detecting gold spectrographically and refers to a number of published papers on the subject.

The standard method for the analysis of gold is the fire assay, and, in view of the modest charge made for gold assays, the prospector would be well advised to send his samples to a reputable assayer rather than attempt "home assays." The main steps in a fire assay are: First, the specimen is fused in a suitable alkali-flux mixture, containing lead oxide and sodium carbonate and/or borax, from which fusion is obtained a lead "button" in which is alloyed all the gold and silver that was in the sample. This is then cupeled by heating in a bone-ash container in an oxidizing atmosphere to eliminate the lead by volatilization and by oxidation and absorption into the body of the container. The bead remaining is made up of the precious metals, from which the silver is extracted by "parting" in nitric acid. The residual gold "sponge" may then be melted down and weighed.

The prospector may in some instances wonder whether a given specimen is worth assaying. The identification service offered by the State Division of Mines and Geology may then be of value. The Division examines at no cost to the sender samples of rocks or minerals that occur in Washington and notifies the sender what the material is. Assays or chemical analyses are not made, but mineralogic and petrographic determinations are. If an assay appears warranted, the sender is so advised, so that he may submit a sample of the material to a commercial assayer if he desires.

---

[15]Scott, W. W., Standard methods of chemical analysis, 5th ed., vol. 1, pp. 431-441, New York, D. Van Nostrand Co., Inc., 1925.

[16]Ahrens, L. H., Spectrochemical analysis. Cambridge, Addison-Wesley Press, Inc., pp. 197-198, 1950.

LOCATING MINING CLAIMS

MINERAL RIGHTS

No license is required to prospect for gold or any other mineral in Washington. The right to mine gold or other minerals on land in Washington may be acquired in one of several ways, depending upon the ownership of the land and its mineral rights. The right is acquired on "open" land simply by making a discovery followed by proper staking and recording, and the mining claim then may be held indefinitely by doing the required $100 worth of assessment work each year. "Open" land is defined and the procedures for obtaining the rights to minerals on land not "open" are described in a leaflet entitled "Mineral Rights," by Sheldon L. Glover, Supervisor, Division of Mines and Geology, and the following is abstracted from that leaflet.

Mining claims, either lode or placer, may be staked without permit or license on the "open" Federal land of the State by any United States citizen or person who in good faith has declared his intention of becoming a citizen. In general, "open" lands include only (1) land of the United States Public Domain (unreserved and unappropriated U. S. public lands) and (2) lands of the National Forests (National Parks are not "open"). The National Forests in Washington are: Olympic, Mount Baker, Snoqualmie, Gifford Pinchot (Columbia), Chelan, Wenatchee, Colville, Kaniksu, and Umatilla. Maps of these National Forests may be obtained from the Regional Forester, U. S. Forest Service, P. O. Box 4137, Portland, Oregon. Records of such Public Domain as remains are kept in the United States Land Office, Federal Building, Spokane 1, Washington, and information about these lands may be obtained there.

Claims may not be staked on land that is not "open." Other procedures, which depend on the ownership of the land, must be followed then if a person desires to prospect or mine on land not belonging to him.

(1) **State-owned land.**—Application should be made to the Commissioner of Public Lands, Public Lands-Social Security Building, Olympia, Washington, for applicable rules and regulations and for a mineral prospecting lease or mining contract on the particular tract, described by legal land description and not to exceed 80 acres. Submerged lands of Puget Sound and of the larger lakes, tidelands of the ocean and sound, and beds of so-called "navigable" streams (many not actually navigable) are either State-owned or (since 1907) have been sold by the State with mineral rights reserved to the State. Information may be obtained from the Commissioner of Public Lands, if there is doubt as to whether land is State-owned or not.

(2) **County-owned land.**—Application should be made to the Board of County Commissoners of the county in which the land is situated for details and permit.

(3) **State Highway rights-of-way.**—Conditions will vary, depending on the status of the land prior to the laying out of the right-of-way. Inquiry may be made of the Director of Highways, Commissioner of Public Lands, or the Division of Mines and Geology.

(4) **Indian Reservations.**—Application should be made to the Agent in charge of the particular reservation for regulations pertaining to prospecting and leasing.

(5) **Privately owned land.**—(Including land owned by any individual or any corporation such as Northern Pacific Railway Co., Weyerhaeuser Timber Co., or other). Application should be made to the owner for permission to trespass, prospect, lease, mine, or purchase. Probably in most instances a private land owner (owner in fee) not only owns the surface rights but also owns the subsurface and all mineral rights to his land. He then may mine or not as he chooses. Anyone else entering the land without the consent of the owner is subject to trespass, he may not stake claims or mine, and he is liable for damages if he persists and injures the land. There are instances, however, where surface rights and mineral rights are separately owned, as described below.

In some land exchanges and in all stock-raising homesteads, all minerals have been reserved to the United States upon issuance of the original patents. In these instances a qualified person may enter the land, prospect for, locate a claim, mine, and remove any of the minerals except those named in the next paragraph. Any person entering such lands to prospect must compensate the owner for any damages caused thereby to crops or improvements.

Ordinarily, the minerals reserved to the United States consist of deposits containing coal, oil, gas, phosphate, sodium, potassium, and oil shale. Lands containing such deposits have not been subject to location since July 17, 1914. Since 1873, coal has been the subject of special laws. Since February 25, 1920, all these minerals may be prospected for under a prospecting permit, and may be leased under certain conditions. Further information about such permits or leases can be obtained from the United States Land Office, Federal Building, Spokane 1, Washington.

Lands containing deposits reserved to the United States are chiefly: (1) lands classified or withdrawn as coal lands and patented thereafter under nonmineral land laws subsequent to March 3, 1909, or (2) lands classified or withdrawn as phosphate, nitrate, potash, oil, gas, and asphaltic minerals, or valuable for such deposits and thereafter patented under the nonmineral land laws (for example, homesteads), subsequent to July 17, 1914.

Whether or not a given tract of land has minerals reserved to the United States, as mentioned above, may be determined by examining the original patent covering the lands in question—this is a matter of record in the office of the County Auditor of the county where the land is situated. Lands which have been patented with mineral reservations are also recorded in the Land Office, at Spokane. The status of all Federal public lands in Washington may be ascertained by correspondence with that office.

In some instances, a previous owner (private individual or a corporation) of a tract of land has reserved to himself the mineral rights when he transferred title to the land. In other words, he may have sold the surface rights and retained ownership of certain or all minerals existing in the land. This has been the case with all State-owned land when the contract of sale was later than 1907, and it is commonly the case with land sold by the Northern Pacific Railway Co. The mineral rights, in such instances, belong to the previous owner of the land, and he has the right of entry for mining purposes, and he may lease his rights or mine if he chooses, but he must compensate the surface owner for damages done to improvements on the land. The only way the surface owner may acquire the mineral rights when they have been reserved by a previous owner is by purchase or lease from the previous owner. Whether or not a previous owner of a given tract of land has reserved the mineral rights when he transferred title to the land may be ascertained by reading the various deeds applicable to the given tract on record in the office of the County Auditor of the county where the land is situated.

Any land owner who may be interested in determining if the mineral rights to his tract were reserved to the United States when patent was issued, or if an owner previous to himself had reserved such rights, may have this information obtained for him by any reliable abstract or title company. The fee for this commonly ranges from a minimum of $10.00 to about $25.00 when many transfer instruments must be searched. Of course, he may search the records himself if he should prefer to do so.

Further information on staking mining claims and on other matters having to do with Federal and State mining laws may be obtained from "An outline of mining laws of the State of Washington" by M. H. Van Nuys (1953).

No list or map has ever been prepared to show (1) lands "open" for mineral location in Washington, (2) claims that have been staked, or (3) State-owned lands that are available for mineral leasing. Information regarding "open" lands and State-owned lands may be obtained from the sources mentioned on page 19. Likewise, no one has published a list of abandoned mining claims. Such a list would be virtually impossible to compile and keep up to date, as well over 125,000 mining claims have been filed in Washington from the earliest record to the end of 1936, and many thousands more have been staked since that time. Records of claim locations and assessment work are not kept by any State or Federal office but are filed with the auditors of the counties in which the claims are located. Information as to the location and current status of previously staked claims can be ascertained only by searching the county records. The auditors also have information as to location and ownership of individual patented claims. Additional information about patented claims may be obtained by examining the original patent survey notes on file in the United States Land Office, Federal Building, Spokane 1, Washington.

The seven full national forests and two fractional parts in Washington cover 9,679,827 acres, which is about 23 percent of the total area of the state. The national forests cover most of the best mineralized areas, thus a large part of the best ground for prospecting is "open" land.

An apparent exception to the general rule that State-owned lands may be leased for mineral exploration and mining are the tidelands (between extreme low- and ordinary high-tide levels) along the shore and beach of the Pacific Ocean from the mouth of the Columbia River north to Cape Flattery, as these tidelands have been set aside by state laws (Chap. 105 and Chap. 110, session laws of 1901, and Chap. 54, session laws of 1935) as a "public highway." These laws forbid the sale or lease of these lands for any purpose (except that tidelands between the mouth of the Queets River and Cape Flattery may be leased for the extraction of petroleum and gas). Later laws, however, allow the leasing of these lands for gas and oil exploration, but apparently no provision has been made for leasing for other minerals such as gold. The land-ownership and mineral-rights situation along the beach and in the lowlands adjacent is further complicated by (1) the fact that the Olympic National Park includes 47,753 acres in a corridor along the Queets River and a strip ½ mile to 3 miles wide back from the beach and extending from the mouth of the Queets River to Cape Alava, near the mouth of the Ozette River, and (2) the presence along the coast of five Indian Reservations—the Quinault, Hoh, Quillayute, Ozette, and Makah Reservations—on some of which it has been established that the mineral rights on the tidelands belong to the Indians.

STAKING LODE CLAIMS

Possessory title to mineral rights on "open" land is established by staking claims, and this entails (1) making a discovery, (2) posting a location notice, (3) staking boundary lines, (4) making a discovery excavation, and (5) recording a location certificate for each claim. A full-size lode claim is 1,500 feet long and 600 feet wide. There is no legal limit to the number of lode claims an individual may hold.

After mineral has been discovered in place the locator should immediately post a location notice at the discovery spot, stating (1) date of discovery (date of posting), (2) name of the claim, (3) name of the locator(s), (4) distance in feet claimed in each direction along the course of the vein from the discovery post, (5) width claimed on each side of the vein, (6) general course of the vein, and (7) description of claim by reference to legal land description or to some natural object or permanent monument.

Within 90 days of the discovery date, and before the claim is recorded, the boundary lines of the claim should be marked by cutting brush and blazing trees along the lines. At each of the four corners of the claim (also at each angle of nonrectangular claims) a substantial stake or stone monument must be erected. A tree or stump will do; if a stake is used, it must be at least 4 inches in diameter; and either a stake or monument must be at least 3 feet high above ground. Each corner must be marked with date of location (same date as recorded on discovery notice) and name of claim. Also, it is recommended that the name of locator(s) and designation of corner (e. g., NW. cor.) be placed on the corner markers. End lines must not be more than 1,500 feet apart, measured along the vein, and to secure full extralateral rights the end lines should be parallel and should cross the vein. The side lines need not be straight, but should follow the course of the vein and should be not more than 300 feet from the center line of the vein.

A discovery excavation must be made within 90 days of the date of discovery and before the claim location notice is recorded. The law relating to discovery shaft requirements (Chap. 45, session laws of 1899, RCW 78.08.130) was amended by the 1955 Legislature—the amendment taking effect June 8, 1955—as follows: "Any open cut, excavation, or tunnel which cuts or exposes a lode and from which a total of two hundred cubic feet of material has been removed, or, in lieu thereof, a test hole drilled on the lode to a minimum depth of twenty feet from the collar, shall hold the lode the same as if a discovery shaft were sunk thereon, and shall be equivalent thereto." (Chap. 357, session laws of 1955.) The amendment also changed another feature of the discovery-shaft law so that a discovery excavation is now required on all mining claims in the state—west of the summit of the Cascade Mountains as well as east of the summit.

After the other steps have been taken, and within 90 days of the date of discovery, the locator must record a copy of his location notice in the office of the auditor of the county in which the lode is found. This notice should contain the same information as is on the location notice on the discovery post.

STAKING PLACER CLAIMS

A placer claim should be in the shape of two or more adjoining 10-acre squares. A claim may be located by one or more persons, but it must be limited in size to 20 acres per locator up to 160 acres for a single placer claim. There is no limit on the number of claims an individual or group may hold. Regardless of the size of a claim (20 acres or 160 acres) only one discovery within it and only $100 annual assessment work is required. On lands covered by a United States survey the placer location should be bounded by lines that conform to the legal survey section lines and section subdivision lines. The claim should be described accordingly (e. g., W½ NE¼NW¼ sec. 22, T. 20 N., R. 17 E.). On lands not covered by a United States survey the claim boundaries must still follow the same rectangular block system used on surveyed lands, and the description must contain reference to natural landmarks or permanent monuments for identification purposes as in lode claims. Exceptions are made allowing the staking of "gulch claims" of irregular shape in the case of placer locations made in narrow gulches with nonmineral sides.

The procedure for staking placer claims is generally like that for lode claims but differs in detail. Placer claims may be staked on "open" lands by (1) making a discovery (a discovery excavation is not required for placer locations), (2) posting notice, (3) staking lines, (4) recording notice, and (5) doing location development work.

At the point of discovery, immediately after the discovery has been made, a notice of location should be posted, giving the name of the claim, name of locator (s), date of discovery (same as date of posting), and description of the claim by reference to legal land survey section subdivisions (or, where the land is not within an area covered by United States survey, by reference to some natural landmark or permanent monument). Within 30 days of the discovery date the locator must mark the boundaries of the claim on the ground in the same manner as in lode locations—by corner stakes at least 4 inches square and 3 feet high or by monuments 3 feet high and by cutting brush and blazing trees along the side and end lines. Also within 30 days of the discovery date the locator must record a copy of the posted location notice in the office of the auditor of the county in which the claim is situated. Within 60 days after discovery he must do upon the claim development work valued at $10 for "each 20 acres or fractional part thereof." This is location work and not to be confused with assessment work. Within a reasonably short time he must record with the same County Auditor an affidavit showing performance of this location work and the nature and kind of work so done.

PATENTING MINING CLAIMS

The owner of a claim held by possessory right has title only to the minerals and such timber on the claim as he uses in his prospecting and mining operations. He may hold these rights only by annually doing assessment work valued at $100. In Washington, taxes are not usually assessed to the owners of such nonpatented claims. Benefits resulting from a patent are: annual assessment work is no longer required; boundaries are established definitely by survey; and the patented claim owner acquires full title to the claim, including land and timber as well as minerals. The patentee may sell or dispose of the timber as he pleases, and may use the claim for any lawful purpose, whether mineral or nonmineral; however, as patented claims are real estate they are subject to taxation as such.

After development work to the value of $500 has been done on a claim the owner may apply for patent to the Bureau of Land Management of the Department of the Interior. A circular outlining procedure for obtaining patents to mining claims is available from the U. S. Land Office, Federal Building, Spokane 1, Washington. The first step is for the claim owner to employ a licensed United States mineral surveyor. He should then request the Area Administrator, Area 1, U. S. Bureau of Land Management, 1001 N. E. Lloyd Boulevard, Portland 14, Oregon, to order the survey to be made. Upon completion of the survey the mineral surveyor's plat and notes must be approved by the Bureau, and then application for patent may be prepared by a qualified attorney and filed in the Land Office.

The recent tendency of the Department of the Interior has been to tighten up on restrictions, and each year fewer and fewer patents are being issued. Two of the things about which the government is strictest are that on each separate claim there must be an adequate discovery and that each claim must be primarily valuable as mineral land. The patent will be denied if the applicant's motive is to acquire land that is valuable primarily for its timber or as a home site, resort site, water power site, as agricultural land, or for other nonmineral use.

The patent costs per claim will vary, depending upon several factors, such as whether only one claim or a group of claims is to be patented. One estimate of the approximate cost for a single lode claim is as follows: surveyor's fee $125 to $250; U. S. Bureau of Land Management office expense $35; attorney's fee $75 to $150; abstract of title $30 to $60; U. S. Land Office entry fee $10; publication fee $30 to $60; purchase price of the land payable to the government (at $5 per acre) $105. Thus the total cost might be between $410 and $670 per claim, but it could well be more.

MINING METHODS AND TREATMENT

Of the United States total production in 1950 of lode and placer gold, 26 percent was recovered by placer methods, 35 percent by amalgamation and cyanidation, and 39 percent by the smelting of ores and concentrates,[17] but of the Washington production of gold in the same year, only 0.04 percent was recovered by placer methods, 0.09 percent by amalgamation, 6.3 percent by cyanidation, 34.5 percent by smelting of ores, and 59.1 percent by smelting of flotation concentrates.[18] Of the placer gold produced in the United States in 1950, 80.6 percent was recovered by bucket-line dredges, and the average value of gravels mined by this method was 15.2 cents per cubic yard. Nonfloating washing plants (dry land dredges) accounted for 14.1 percent of the placer production, operating on gravels averaging 35.3 cents per cubic yard; dragline dredges, 3.4 percent, on 15.9-cent gravel; underground (drift) placers, 0.1 percent, on $2.195-gravel; hydraulicking, 0.7 percent, on 23.8-cent gravel; small-scale hand methods (wet), 0.8 percent, on 65-cent gravel; small-scale hand methods (dry), 0.01 percent, on $1.40-gravel; and suction dredges, 0.2 percent, on 18.9-cent gravel.[19]

A recently published report[20] describes equipment and methods useful to prospectors and operators of small placers, and both small-scale and large-scale placer mining methods have been described

[17]Bell, J. E., Gold and silver: U. S. Bur. Mines Minerals Yearbook, p. 558, 1950.

[18]Robertson, A. F., and Halverson, Virginia, Washington gold, silver, lead, and zinc: U. S. Bur. Mines Minerals Yearbook, p. 1608, 1950.

[19]Bell, J. E., op. cit., pp. 578-580, 1950.

[20]Masson, D. L., Small scale placer mining: Washington State Inst. Tech. Bull. 45 M, 1953.

by Averill.[21] Details of the various mining methods used in both placer and lode-gold mines are given in many textbooks and handbooks,[22] in publications of the American Institute of Mining and Metallurgical Engineers, and in reports of the U. S. Bureau of Mines. Of the Bureau of Mines reports, a bulletin dealing with metal-mining practice in general[23] is especially helpful, and two information circulars describe mining methods and costs at two of this state's leading gold producers, the Knob Hill mine[24] and the Holden mine.[25] For other citations to published articles on mining and milling practices at these and other gold-producing mines in the state the reader may refer to the descriptions of the individual properties on pages 39 to 134.

Because gold is present in practically unlimited quantity (thousands of millions of tons) in the oceans of the world, many processes, some meritorious and some fraudulent, have been proposed for extracting gold from this source, but none has proved commercially feasible. One attempt to recover gold from sea water was a Washington operation that was active from the middle to the late 1930's. The following information is based on statements made by the inventor and observations of disinterested parties. No statements of fact or opinion regarding the process are to be attributed to the Division of Mines and Geology. The method of recovery was based on the collection of gold—supposedly in colloidal form in the sea water—on amalgamated zinc shavings. It was proved by the inventor that satisfactory results could be repeatedly obtained in a laboratory-scale pilot plant using one ton of sea water in a batch process, but the operation of a later commercial-scale plant, built on a pier extending over the water of Puget Sound at Lakota, King County, was not successful due to mechanical and other operational difficulties involving, among other things, inability to control and maintain certain precise and vital technical adjustments. The problems included precise pH adjustments and some galvanic phenomena that defied control in the large-scale, continuous-flow operation but could be maintained in the one-ton pilot model, where most encouraging recoveries of both gold and silver were obtained repeatedly. More than 500 assays on samples of the pilot-plant product indicated that the gold-silver recovery was high enough to justify a commercial operation if it could duplicate the pilot-plant recovery.

[21]Averill, C. V., Placer mining for gold in California: California Div. Mines Bull. 135, pp. 11-144, 1946.

[22]Peele, Robert, Mining engineers' handbook, 3d ed., New York, John Wiley & Sons, Inc., 1941.

[23]Jackson, C. F., and Hedges, J. H., Metal-mining practice: U. S. Bur. Mines Bull. 419, 1939.

[24]Cooper, C. L., Mining and milling methods and costs at Knob Hill mine, Republic, Washington: U. S. Bur. Mines Inf. Circ. 7123, 1940.

[25]Soderberg, R. L., Mining methods of the Holden mine, Howe Sound Co., Chelan Division, Holden, Washington: U. S. Bur. Mines Inf. Circ. 7448, 1948.

MARKET AND PRICES

Since January 3, 1934 the domestic gold market has been governed by the United States Gold Reserve Act of 1934, which is summarized by Van Nuys[26] as follows:

Under this Act the United States went off the gold standard, withdrew gold coins from circulation, and authorized the Secretary of the Treasury to make regulations governing the acquisition, holding, melting and treating, and the importing and exporting of gold. These regulations require special licenses from the Secretary for doing of any of the acts just enumerated. Such licenses are difficult to obtain. So far as the Act affects gold mining, an important exception relates to "gold in its natural state"; viz., "gold recovered from natural sources which has not been melted, smelted, or refined or otherwise treated by heating or by a chemical or electrical process." Such kind of gold may be "acquired, transported within the United States, imported, or held in custody for domestic account without the necessity of holding a license therefor." Accordingly, a miner (usually placer) who produces gold in such natural state may hold it, import it (if produced outside the United States), transport it within the United States, and sell it, without a license; and his purchaser and assigns may do likewise wihout a license. But no one, not even the miner, may export such gold or any gold without a license.

The regulations permit the miner, without a license, to [treat] gold amalgam by heating the natural gold sufficiently to separate the mercury without melting the gold; the total retort sponge held by him at any one time is not to exceed 200 troy ounces. The miner may sell such sponge without a license, but only to a United States Mint or to a private party having a license. When the miner sells his natural gold or sponge to a private party, usually a bank or a dealer, or to a U. S. Mint, the miner must furnish the purchaser with a written certificate, Form TG-19, showing he (miner) mined it, etc.; and the purchaser (when he sells it) must likewise furnish his purchaser in turn with a certificate, Form TG-21, showing from whom he purchased, etc.

The U. S. Mints (in 1953) pay $35.00 for an ounce of fine gold, less ¼ of 1 percent and less Mint charges, subject to change by the Secretary. The U. S. Mint (Assay Office) at Seattle[27] serves Washington, Idaho, Montana, Oregon, and Alaska. "United States Mint" means a United States Mint or Assay Office.

The value of gold has increased gradually for more than 2,000 years. The price was a little over $20 per fine troy ounce for more than 200 years until January 1934, when the President of the United States set the U. S. Mint price at $35 per ounce, at which it has remained. A detailed chronology of events leading to the revaluation of gold in 1934 is given by Libbey and Mason.[28] Henderson[29] has summarized the monetary history of gold in this country as far back as 1786.

---

[26]Van Nuys, M. H., An outline of mining laws of the State of Washington: Washington Div. Mines and Geology Bull. 41, pp. 125-126, 1953.

[27]The U. S. Mint (Assay Office) in Seattle was closed and, as of February 28, 1955, its gold purchasing functions were transferred to the U. S. Mint, San Francisco, California and the U. S. Mint, Denver, Colorado.

[28]Libbey, F. W., and Mason, R. S., Chronology of events leading to revaluation of gold in 1934: Oregon Dept. Geol. and Min. Ind., The Ore.-Bin, vol. 16, no. 2, p. 709, 1954.

[29]Henderson, C. W., Gold and silver: U. S. Bur. Mines Minerals Yearbook, 1932-1933, pp. 11-17, 1933.

In 1948 and 1949 much publicity was given to the possibility of establishing a premium price for "natural" gold on the open market among hoarders preferring gold to currency and among speculators anticipating a rise in the Mint price for gold. It was estimated[30] that "natural" domestic gold containing 25,000 ounces of fine gold was sold at prices up to $43 per ounce in 1948, and 29,000 ounces at an average price of $39 to $40 in 1949. Such sales declined in volume in 1950, perhaps reflecting pessimism regarding a possible rise in the Mint price. In the past few years gold has been quoted at from $35 to more than $50 per ounce on the free markets in various countries, but much of the trade is conducted secretly in black markets, so that exact prices paid are difficult to determine.

HISTORY OF GOLD IN WASHINGTON

The first discovery of gold in Washington was probably that reported by Stevens,[31] who, in describing explorations made by Captain McClellan and his party in search of a railroad route through the Cascades in 1853, said, "Here the first traces of gold were discovered, and though not sufficiently abundant to pay for working, it caused considerable excitement in the camp." The locality referred to was in the Yakima Valley, though it is not clear whether it was near the mouth of the Naches River or was on the headwaters of the Yakima River. In the same report Stevens[32] wrote, "It is also worthy of observation that gold was found to exist, in the explorations of 1853, throughout the whole region between the Cascades and the main Columbia to north of the boundary, and paying localities have since been found at several points, particularly on the southern tributary [probably Peshastin Creek] of the Wenatshapan [Wenatchee River]. The gold quartz also is found on the Nachess River."

In 1855 there was a small gold rush to the Colville region,[33] and on September 14 of that year the Oregonian newspaper reported prospectors recovering $5 to $8 per day using rockers along the Pend Oreille River. On September 28 and November 23 the Olympia Pioneer and Democrat newspaper reported placer-gold discoveries along the Columbia River from the mouth of the Spokane River to the mouth of the Pend Oreille River and up that river for at least 40 miles. Prospectors were said to be making fairly good recoveries on Sheep Creek just south of the Canadian boundary. The placers in this region appear to have been too small and too low grade to hold the prospectors' interest for long, and their attention was soon directed to other areas.

[30]Bell, J. E., Gold and silver: U. S. Bur. Mines Minerals Yearbook, p. 561, 1950.

[31]Stevens, I. I., Explorations for a railroad route from the Mississippi River to the Pacific Ocean: Pacific Railway Reports, vol. 12, no. 1, p. 140, 1860.

[32]Idem, p. 257.

[33]Trimble, W. J., The mining advance into the Inland Empire: Univ. Wisconsin Bull. 638, History Ser., vol. 3, no. 2, pp. 15-17, 1914.

From the beginning, gold-mining activities were intermittent rather than constant. Periods of activity were followed by periods of stagnation. These cycles were influenced to some extent by economic conditions. In general, gold mining has fared relatively well and prospecting has flourished during economic recessions, and gold mining has been at a disadvantage and has declined during periods of general prosperity and high prices. Of even more influence locally has been the prospectors' urge to move on to possibly more lucrative fields as news was spread of gold strikes in new areas. Thus, though a party that was surveying the international boundary discovered gold on the Similkameen River in Okanogan County in 1859,[34] the men who rushed to the Similkameen stayed only about three months and then flocked northward to the Frazier River and Cariboo district, leaving the Similkameen placers practically deserted. A few months later the surge was reversed and many of the same men returned to the Similkameen, and others spread out in both eastern and western Washington. Hodges[35] reports that these men returning from the north discovered placer gold on Ruby Creek in Whatcom County, on the Sultan River in Snohomish County, and on Peshastin and Swauk Creeks in Chelan and Kittitas Counties in the early 1860's, but Bethune[36] sets the dates for discovery of gold in these districts at later dates, as follows: Swauk Creek in 1874, Peshastin Creek in 1877, Ruby Creek in 1878, and Cle Elum River in 1881. Bethune [37] records placer mining on Cassimer Bar at the mouth of the Okanogan River as early as 1860 and increased placer activities along this river in the middle 1880's.

Bars along the Columbia River adjacent to the Colville Indian Reservation were worked intermittently, mainly by Chinese, from 1870 to 1890,[38] and Columbia River bars above Priest Rapids were reported[39] to have been worked prior to 1860. Other bars of the Columbia, Snake, and most of the other rivers in the state were prospected, and some of these yielded moderate amounts of placer gold. Beach placers were discovered along the Pacific Coast and were worked in a small way for many years. During a brief gold excitement in 1894 the beaches were staked for 60 or 70 miles south of Cape Flattery, but productive localities were found to be limited to 20 miles south of Portage Head.[40] By 1900 most of the state's placers had been found and largely worked out.

[34]Bethune, G. A., Mines and minerals of Washington: Washington State Geologist 1st Ann. Rept., pp. 6-7, 1890.

[35]Hodges, L. K., Mining in the Pacific Northwest: Seattle Post-Intelligencer, pp. 6-7, 1897.

[36]Bethune, G. A., op. cit., pp. 8-12, 1890.

[37]Idem, pp. 9-10.

[38]Pardee, J. T., Geology and mineral resources of the Colville Indian Reservation, Washington: U. S. Geol. Survey Bull. 677, p. 53, 1918.

[39]Browne, J. R., Resources of the Pacific slope, New York, D. Appleton and Co., p. 568, 1869.

[40]Arnold, Ralph, Gold placers of the coast of Washington: U. S. Geol. Survey Bull. 260, pp. 154-157, 1905.

Nearly all the lode mining districts in the state have some mines that produce gold, at least as a byproduct, and most of these districts had been discovered by 1897, when a historically interesting account of the mines of the Pacific Northwest was published.[41] One of the first discoveries of lode gold in the state was at the base of Mount Chopaka in 1871. In 1878 C. P. Culver discovered lode gold in the Blewett district, where the first stamp mill was built in the then Territory of Washington.[42] Lode gold was discovered at Monte Cristo in Snohomish County in 1889 and at Republic in Ferry County in 1896.[43] Previous to 1898 gold quartz veins were being mined and the ore milled in arrastres near Liberty in the Swauk district of Kittitas County.[44] A few very early discoveries of lode gold were made within the area of the Colville Indian Reservation, but it was not until the Reservation was opened to mineral entry in 1898 that mining began to expand in this region. The Republic district has been the leading lode gold camp in the state, but other important gold-producing areas have been the Railroad Creek, Blewett, Mount Baker, Monte Cristo, Slate Creek, Oroville-Nighthawk, Orient, and Wenatchee districts.

Lode gold has been mined or reported in 23 counties, and placer gold in 25 counties in Washington. The following list includes most of the placer gold localities in the state. At some of these localities only colors have been found; at others small amounts of gold are reported to have been recovered; and at a few of the places listed fairly large profitable operations have been carried on.

PLACER GOLD LOCALITIES

ASOTIN COUNTY
  Snake River bars

BENTON COUNTY
  Columbia River bars

CHELAN COUNTY
  Blewett area
  Bridge Creek
  Chiwawa River
  Columbia River bars
  Entiat River
  Lakeside (Lake Chelan)
  Leavenworth
  Nigger Creek
  Peshastin Creek
  Railroad Creek
  Wenatchee River

CLALLAM COUNTY
  Cedar Creek
  Ozette Beach
  Sand Point
  Shi Shi Beach
  Yellow Banks

CLARK COUNTY
  South Fork Lewis River

DOUGLAS COUNTY
  Columbia River bars

FERRY COUNTY
  Bridge Creek
  Columbia River bars
  Covada district

[41]Hodges, L. K., op. cit., 1897.

[42]Bethune, G. A., op. cit., pp. 10-12, 1890.

[43]Fischer, A. H., A summary of mining in the state of Washington: Univ. Washington Eng. Exper. Sta. Bull. 4, pp. 18-21, 1918.

[44]Russell, I. C., A preliminary paper on the geology of the Cascade Mountains in northern Washington: U. S. Geol. Survey Ann. Rept. 20, pt. 2, p. 207, 1900.

PLACER GOLD LOCALITIES—*Continued*

FERRY COUNTY—*Continued*
Danville area
Kettle River bars
Republic
Sanpoil River

GARFIELD COUNTY
Snake River

GRANT COUNTY
Columbia River bars

GRAYS HARBOR COUNTY
Cow Point
Damons Point
Moclips River
Point Brown

JEFFERSON COUNTY
Ruby Beach

KING COUNTY
Denny Creek
Money Creek
Tolt River

KITTITAS COUNTY
Cle Elum River
Columbia River bars
Liberty
Manastash Creek
Naneum Creek
Swauk Creek
Teanaway River
Yakima River

LINCOLN COUNTY
Columbia River bars

OKANOGAN COUNTY
Columbia River bars
Mary Ann Creek
Methow River
Myers Creek
Nighthawk
Oroville
Park City
Similkameen River
Squaw Creek
Twisp
Wauconda

PACIFIC COUNTY
Fort Canby

PEND OREILLE COUNTY
Pend Oreille River bars
Russian Creek
Sullivan Creek

PIERCE COUNTY
Silver Creek

SKAGIT COUNTY
Skagit River

SKAMANIA COUNTY
Camp Creek
McCoy Creek
Texas Gulch

SNOHOMISH COUNTY
Darrington area
Granite Falls
Sultan River

STEVENS COUNTY
Columbia River bars
Kettle Falls
Kettle River bars
Marcus
Meyers Falls
Northport district
Orient

WHATCOM COUNTY
Acme vicinity
Mount Baker
Ruby Creek
Silesia Creek
Slate Creek

WHITMAN COUNTY
Snake River bars

YAKIMA COUNTY
American River
Morse Creek
Summit district
Surveyors Creek

PRODUCTION

Gold production forms an important part of the total mineral output of the state of Washington. Every year prior to 1916 gold outranked in value all other metals produced, and it has ranked first in many of the years since that date, but in 1952 gold accounted for only 13 percent of the value of metallic production in the state. Annual gold production in Washington from 1866 through 1902 is shown in the table on page 33, and annual totals for the state since that date, as well as data for individual counties, are given in the table on pages 34 and 35.

The following data on production, except as otherwise noted, are from the annual volumes of Mineral Resources and its successor, the Minerals Yearbook, published prior to 1924 by the United States Geological Survey and later by the United States Bureau of Mines. Total Washington state gold production from 1860 through 1952 was 2,570,002 ounces, valued at $68,705,393, which ranks Washington eleventh among the states as a gold producer. Total United States gold output from the beginning of the record through 1951 has been 283,276,495 ounces, and the Washington share of this amount has been slightly less than 1 percent. The greatest production of gold in Washington in any single year since 1900 was 92,117 ounces, valued at $3,224,095, in 1950. In that year 97 percent of the state production came from three mines: the Holden, Gold King, and Knob Hill, which ranked in production twelfth, fourteenth, and eighteenth, respectively, among the gold mines of the United States. In the next year, 1951, the Gold King mine was the leading producer in this state, ranking eleventh nationally, followed by the Holden mine (thirteenth) and the Knob Hill mine (sixteenth). Combined production from the Holden and Gold King mines in Chelan County outranked that of all but six other gold districts in the United States in 1951. Prior to 1900 most of the gold produced in the state was from placers, but since that time less than 2 percent of the output has been from placers. Placer production from 1900 through 1953 was $930,597. To illustrate further, placer production for the 10-year period from 1900 through 1909 was $350,541 but for a similar period from 1944 through 1953 was only $12,456.

The Republic district of Ferry County has had the longest consistent record of gold production, and Ferry County has led all counties during 28 of the 50 years from 1903 to 1952, inclusive. Chelan County, formerly a small producer, took the lead in 1938 as a result of the output of the Holden mine and continued to hold it (except for 1947) up through 1953. Stevens County produced the greatest amount of gold during 5 years (1905 through 1908, and 1922), and Whatcom County led for 3 years (1904, 1929, and 1930). Okanogan, Snohomish, and Kittitas Counties have nearly always produced some gold, and in some years important amounts. In total gold production for the years 1903 through 1952, Chelan and Ferry Counties are at the top of the list with about $21 million each, followed in order of output by Whatcom, Stevens, Okanogan, Snohomish, and Kittitas Counties.

Gold Production in Washington prior to 1903

Year	Fine ounces	Value
Prior to 1866		$9,000,000
1866		1,000,000
1867		400,000
1868		400,000
1869		300,000
1870		300,000
1871		320,107
1872		260,000
1873		206,341
1874		154,535
1875		81,932
1876		26,988
1877		300,000
1878		300,000
1879		75,000
1880		135,800
1881	5,806	120,000
1882	5,805	120,000
1883	3,870	80,000
1884	4,112	73,952
1885	5,805	126,172
1886	7,111	147,548
1887	7,257	160,503
1888	7,015	145,000
1889	9,001	193,709
1890	9,869	204,000
1891	16,206	371,897
1892	18,071	373,553
1893	11,049	228,394
1894	11,261	232,761
1895	18,053	373,148
1896	19,134	395,490
1897	21,754	449,664
1898	37,065	612,118
1899	33,156	729,388
1900	34,743	732,437
1901	28,082	661,240
1902	13,166	374,471
Total	327,391	$20,166,148

Early statistics are not as complete as might be desired, and some early estimates of production appear to have been too high. Total gold and silver production in Washington through 1866 was estimated by J. Ross Browne,[45] United States Commissioner of Mining Statistics, to be $10,000,000. Estimates of early gold production from a few districts follow: $1,500,000 (placer and lode) from the Blewett district from 1870 through 1900, and $200,000 from 1901 through 1910;[46] $500,000 from the Similkameen placers up to

---

[45]Browne, J. R., Resources of the Pacific slope, New York, D. Appleton and Co., p. 6, 1869.

[46]Weaver, C. E., Geology and ore deposits of the Blewett mining district: Washington Geol. Survey Bull. 6, p. 71, 1911.

Gold Production

Year	Asotin	Benton	Chelan	Clallam	Douglas	Ferry	Grant	King	Kittitas	Lincoln
1903....	$690	....	$80,090	....	....	$275,397	....	$1,584	$4,434	$36
1904....	....	....	(a)21,187	....	....	113,257	....	6,327	2,675	....
1905....	1,025	....	12,640	....	....	86,053	....	(b)18,657	....	....
1906....	983	....	290	....	....	70,742	....	(c)21,828	....	....
1907....	....	....	2,766	....	....	4,032	....	10,345	8,686	....
1908....	....	$21	3,630	$3,614	....	17,333	....	12,645	5,971	....
1909....	....	....	4,232	39	....	210,437	....	10,501	5,829	....
1910....	68	....	6,867	....	....	714,808	....	19,420	3,589	....
1911....	....	....	21,846	....	....	778,526	....	11,245	2,081	4
1912....	....	....	35,130	256	....	605,698	....	7,711	2,141	....
1913....	....	....	8,590	....	....	645,009	....	4,568	3,677	....
1914....	....	....	1,575	110	....	513,276	....	....	4,387	....
1915....	....	166	4,279	685	....	351,973	....	....	3,990	....
1916....	....	....	8,682	372	....	399,376	....	1,344	4,832	....
1917....	....	....	683	624	....	332,071	....	5,742	4,848	....
1918....	....	....	....	239	....	276,066	....	4,674	2,635	....
1919....	....	....	135	....	....	245,141	....	1,856	973	....
1920....	....	....	27	257	....	110,278	....	1,137	1,075	....
1921....	....	....	207	....	....	88,290	....	....	2,845	....
1922....	57	....	169	180	....	85,686	....	7	2,628	....
1923....	....	....	....	....	....	276,605	....	....	1,240	....
1924....	....	....	....	....	....	280,751	....	....	625	....
1925....	....	....	81	....	....	133,809	$52	....	429	....
1926....	....	....	....	94	....	156,802	....	1,798	27,480	....
1927....	....	....	424	....	....	311,805	....	785	389	....
1928....	....	....	....	....	....	258,793	....	....	1,878	....
1929....	....	....	21	....	....	10,551	....	....	6,073	....
1930....	....	....	50	....	....	9,464	....	....	1,143	....
1931....	32	....	831	....	....	40,706	....	....	1,820	....
1932(d)..										
1933....	1,690	192	1,929	1,150	$88	63,894	89	48	6,154	3,710
1934....	4,467	1,309	3,031	857	187	161,491	150	361	7,518	8,504
1935....	1,729	1,554	2,541	1,533	1,365	177,933	406	....	9,681	1,736
1936....	1,526	490	1,176	308	588	245,203	378	1,589	6,405	56
1937....	910	700	41,265	....	35	642,740	455	1,190	3,465	....
1938....	3,745	1,190	1,103,690	315	1,260	906,430	805	6,440	2,975	455
1939....	5,985	1,085	1,697,500	455	3,605	1,025,045	2,975	6,825	2,415	910
1940....	4,025	840	1,810,235	280	1,260	781,725	1,610	1,330	56,665	....
1941....	4,515	945	1,611,680	....	70	1,209,565	1,330	3,885	665	....
1942....	6,055	3,640	1,703,975	....	35	883,960	2,485	....	105	....
1943....	70	3,990	1,467,200	....	....	802,445	....	....	....	....
1944....	....	1,925	921,445	....	....	716,765	....	....	105	....
1945....	....	385	1,415,750	....	....	667,405	....	....	70	....
1946....	2,975	35	1,136,660	....	....	649,705	....	....	385	....
1947....	490	210	422,485	....	....	790,650	....	....	....	....
1948....	....	....	1,464,155	....	....	986,860	....	....	....	....
1949....	....	35	1,686,405	....	....	831,285	....	....	70	....
1950....	35	....	2,264,885	....	....	872,515	....	....	2,660	....
1951....	....	....	1,626,030	....	....	(e)683,865	....	70	....	....
1952....	....	....	(f)1,894,725	....	....	....	....	35	1,855	....
Total...	$41,072	$18,712	$22,491,194	$7,754	$8,493	$20,442,521	$10,735	$163,947	$209,566	$15,411

(a) Includes production of Asotin and Clark Counties in 1904.
(b) Includes production of Franklin and Kittitas Counties in 1905.
(c) Includes production of Kittitas County in 1906.
(d) Production figures by counties not available.
(e) Includes production of Kittitas and Whatcom Counties in 1951.
(f) Includes production of Ferry County in 1952.
(g) Includes production of Stevens County in 1952.

By Counties

Okanogan	PendOreille	Skamania	Snohomish	Stevens	Whatcom	Whitman	Yakima	Others	Total
$36,009	....	....	$70,661	$2,502	$36,388	....	....	①②	$507,885
18,066			26,351	11,600	115,000				314,463
2,313			40,425	165,863	77,983			③	405,048
8,799			2,873	77,837	38,196			③	221,648
9,394			45,504	149,588	28,759				259,047
7,998			775	183,893	6,045	$309			242,234
8,907			49	121,498	559				362,051
13,088			88	30,182	26			①②	788,145
4,723	$633		433	27,074	189	203			847,677
3,302			717	23,823	2,186				680,964
2,837			2,622	23,480	5,492				696,275
2,724			7,217	11,894	15,635				557,173
10,281			3,462	4,063	12,520	355			391,419
18,992			1,844	8,972	133,200			①	577,655
6,403	170		518	3,757	137,382		$34	②④	492,324
1,586	3		1,269	8,199	9,987				304,658
1,389	100		1,795	1,434	....			⑤	252,862
2,553			3,010	2,004	519				120,860
1,923			198	580	34,443				128,486
1,428	5		107	1,019	95,679				186,965
1,623	37		1,178	1,373	60,011				342,067
4,646			2,081	2,696	18,745			⑥	309,617
258	35		3,033	1,583	90,800			①⑤	230,253
379	95		3,948	2,438	58				193,092
....	228		2,749	157	86,822			①	403,380
918	21		2,336	810	72,404			②	337,167
103	166		3,808	841	55,274			⑦	76,898
2,483	197		1,943	724	71,744				87,748
1,697	324		219	1,875	12,465	66			60,035
....									105,057
6,820	385	....	1,512	3,204	2,334	782	291	⑦	94,319
13,785	438	$290	1,757	83,050	1,011	281	457	⑥⑧⑨	290,149
56,903	210	560	2,877	61,061	18,809	1,365	126	②③⑧⑨	340,886
125,979	112	140	1,533	9,170	31,003	1,008	756	⑨	427,609
86,695	350	175	805	29,995	461,895	175	....		1,270,850
43,330	245	875	1,365	91,700	430,290	945	35	⑨	2,596,125
270,120	280	280	1,890	33,845	111,265	175		⑧	3,164,700
144,305	....	210	6,930	52,990	12,075	280	....		2,874,760
36,260	105		4,375	59,080	13,230	315	35	①	2,946,160
10,500			1,260	24,920	1,925				2,638,860
70			385	9,345	35				2,283,340
910			140	13,370	....		35		1,654,695
770	35		140	175				①	2,025,100
35			140	840	....		105		1,790,880
2,345			35	3,850	3,675			⑦	1,223,775
350			70	980	210				2,452,625
....			455	735	805				2,519,790
82,355			35	525	....			⑦	3,224,095
47,600	1,050		....	105				⑦	2,359,175
16,905	(g)1,085		35	....	2,520				1,917,160
$1,111,705	$6,309	$2,530	$256,952	$1,350,699	$2,309,593	$6,364	$1,769	$4,564	$48,578,436

① Pierce County: 1903—$50, 1910—$8, 1916—$41, 1925—$156, 1927—$21, 1941—$105, 1945—$70. Total—$451.
② Skagit County: 1903—$44, 1910—$1, 1917—$12, 1928—$2, 1935—$35. Total—$94.
③ Clark County: 1905—$100, 1906—$100, 1935—$182. Total—$382.
④ Wahkiakum County: 1917—$80. Total—$80.
⑤ Lewis County: 1919—$39, 1925—$17. Total—$56.
⑥ Walla Walla County: 1924—$73, 1934—$237. Total—$310.
⑦ Garfield County: 1929—$61, 1933—$47, 1947—$35, 1950—$1,085, 1951—$455. Total—$1,683.
⑧ Columbia County: 1934—$110, 1935—$98, 1939—$35. Total—$243.
⑨ Grays Harbor County: 1934—$858, 1935—$182, 1936—$189, 1938—$35. Total—$1,264.

1911;[47] $1,800,000 from the Republic district through 1910;[48] and about $7,000,000 from the Monte Cristo district through 1918.[49]

OCCURRENCES

GENERAL STATEMENT

The following lists of lode and placer gold occurrences are extracted and condensed from similar but more complete lists in Division of Mines and Geology Bulletin 37, Inventory of Washington minerals, Part 2, Metallic minerals, a report that is in preparation in 1955, but the publication date of which is yet undetermined. The original lists in Bulletin 37 are as complete as it is possible to make them—including every occurrence known to the Division. Those complete inventories include many properties where gold is present in trace or only very minor amounts along with other metals which are the principal constituents of the ore. Most occurrences where gold is of secondary or minor value are not included in the following pages, and anyone wanting the complete list may refer to Bulletin 37, Part 2. Although most of the following occurrences have either been in production in the past or are considered to merit further prospecting, a few of the occurrences shown do not fall in either of these categories, but are included for some special reason— (1) to illustrate the geographic distribution of gold deposits in the state, (2) because of special interest in gold in certain areas even though probabilities of profitable operations in those areas are very slight, or (3) because of special interest or considerable publicity given a few individual deposits of little or no probable value.

The lode gold occurrences are described first. These descriptions are followed by maps (facing pp. 111 and 112) showing the locations of both the lode and placer occurrences. Then follow the descriptions of placer gold deposits. The properties are arranged by counties and are listed within the counties alphabetically by name. The numbers following the names correspond to the numbers designating those same properties on the maps.

To facilitate tracing a numbered symbol on a map to the description of the property represented by the symbol, a finding list (numerically arranged) relating number to property name is included with each map. The scale of the maps is such that they can be regarded as index maps only. Locations of the properties are given as precisely as possible in the text, but, because of the small

[47]Umpleby, J. B., Geology and ore deposits of the Oroville-Nighthawk mining district: Washington Geol. Survey Bull. 5, pt. 2, p. 76, 1911.

[48]Umpleby, J. B., Geology and ore deposits of the Republic mining district: Washington Geol. Survey Bull. 1, p. 34, 1910.

[49]Fischer, A. H., A summary of mining in the state of Washington: Univ. Washington Eng. Exper. Sta. Bull. 4, p. 20, 1918.

map scale, the detail is not as great as it would be on larger scale maps. In order to indicate the locations of all the occurrences in some areas it is necessary to allow a single symbol to represent several closely spaced properties.

The property descriptions from which these lists are condensed were compiled from a wide variety of sources, of varying degrees of reliability. Thus, the critical reader probably may find inaccuracies and may recognize that some properties are inadequately described. Little information is available on most of the placer deposits, reflecting the comparatively minor value of the state's placers.

For the reader's convenience the occurrences are described under a standardized set of 14 headings: **Loc** (location), **Elev** (elevation), **Access, Prop** (property), **Owner, Ore, Ore min** (ore minerals), **Gangue, Deposit, Dev** (development), **Improv** (improvements), **Assays, Prod** (production), and **Ref** (references). A date in parentheses following any information in the descriptions indicates the date at which that information was assumed to be correct. Legal land descriptions are abbreviated; thus, sec. 3, (40-25E) indicates section 3, Township 40 North, Range 25 East, Willamette Meridian. Under the heading **Ref** is a list of abbreviated references to the sources of information for each property. The reference is made by a number (in boldface type) which refers to a title under the same number in the bibliography for occurrences on pages 135 to 138. This is followed by the page reference or, in the case of a periodical, by a date (written 7/5/34 [July 5, 1934], or 5/34 [July 1934], or simply 1934), and then the page reference. Where the reference includes several issues of a periodical, the parts of the reference are separated by semicolons. Thus **1**, 4/18, p. 19; 7/18, pp. 45-46, refers to Alaska and Northwest Mining Journal for April 1918, page 19, and August 1918, pages 45 to 46. Other abbreviations used in the property descriptions are in the following list.

ABBREVIATIONS

Ag—silver
approx.—approximately
As—arsenic
Au—gold
av.—average, averaged, averaging
Ave.—avenue
Bi—bismuth
Bros.—Brothers
Bur.—Bureau
C.—centigrade
Co.—Company
conc.—concentrate, concentrates
cor.—corner
Corp.—Corporation
Cr—chromium
Cr.—Creek
Cu—copper
cu.—cubic
Dept.—Department
Dev—development
dia.—diameter
dist.—district
Div.—Division
E.—east
Elev—elevation
est.—estimated
et al.—et alii (and others)
F.—Fahrenheit
Fe—iron
Fk.—Fork
fr.—fraction
ft.—foot, feet
gm.—gram, grams
Hg—mercury
Improv—improvements
in.—inch, inches
Inc.—Incorporated
Is.—Island
lb.—pound, pounds
Lk., lk.—Lake, lake
Loc—location
Ltd.—Limited

max.—maximum
mi.—mile, miles
min.—minimum
Mt.—Mount
Mtn.—Mountain
N.—north
NE.—northeast
Ni—nickel
no.—number
Nos.—Numbers
NW.—northwest
O—oxygen
Ore min—ore mineral(s)
oz.—ounce, ounces
p.—page
Pb—lead
%—percent
pp.—pages
Prod—production
Prop—property
Pt—platinum
R.—river, range
Ref—references
Ry.—railway
S.—south
Sb—antimony
Se—selenium
SE.—southeast
sec.—section
sq.—square
SW.—southwest
T.—township
tr.—trace
U—uranium
U. S.—United States
vol.—volume
W—tungsten
W.—west
Wash.—Washington
yd.—yard, yards
yr.—year, years
Zn—zinc

LODE GOLD OCCURRENCES

BENTON COUNTY

Prosser (165)

Loc: SW¼ sec. 33, (9-25E), about 3 mi. E. of Prosser on S. side of the highway. **Access:** Highway. **Owner:** R. W. Wilson, Grandview, Wash. (1934). **Ore:** Gold, silver. **Ore min:** Pyrite. **Deposit:** Mineralized zone along contact of basalt with interbasalt sediments. **Dev:** Shaft 50 ft. deep and open cut 100 ft. long. **Assays:** 26¢ to $1.60 Au and about the same values in Ag. One assay showed $13.00 Au (1934). **Ref:** 104, 6/30/36, p. 24. **158.**

CHELAN COUNTY

Alta Vista (Pole Pick No. 2) (127)

Loc: SW¼ sec. 2, (22-17E), on S. side of Culver Gulch. **Elev:** 3,700 ft. **Access:** Road and trail. **Prop:** 1 claim: Pole Pick No. 2. **Owner:** Alta Vista Mining Co. (1911). **Ore:** Gold. **Ore min:** Arsenopyrite, pyrite. **Gangue:** Quartz, calcite, talc. **Deposit:** 4-ft. quartz-calcite vein in peridotite. The vein is heavily impregnated with ore minerals. **Dev:** Total of about 1,190 ft. of cross-cuts and drifts. **Assays:** Ranged from 0.04 oz. to 1.6 oz. Au. Main vein ranged from $8 to $200 Au. **Prod:** Amount not known. **Ref:** 63, p. 73. **161**, pp. 89-91.

Black and White (Diamond Dick) (128)

Loc: SW¼ sec. 12, (22-17E), on E. side of Peshastin Cr. several hundred yards S. of Blewett. **Access:** 50 ft. above highway U. S. 97. **Prop:** 2 claims (1953). **Owner:** Wm. Johnson, Wenatchee, Wash. (1953 ——). H. Whitely and John White (1938). **Ore:** Gold, silver. **Ore. min:** Free gold, pyrite, chalcopyrite, arsenopyrite. **Gangue:** Quartz, calcite. **Deposit:** Small ore-bearing stringers of quartz and calcite in a shear zone in serpentine. **Dev:** A shaft and 3 adits with raises and stopes, totaling at least 850 ft. of workings. **Improv:** 40-ton mill in good condition (1942). **Assays:** Reported $8 Au, 80¢ Ag. **Prod:** Small amounts in 1934 and 1939. **Ref:** 67, p. 9. 97, 1935, p. 352; 1940, p. 476. **158.**

Black Jack (Blewett, La Rica) (129)
(See also Gold Bond.)

Loc: SW¼ sec. 1, (22-17E), near Peshastin Cr., Blewett dist. **Elev:** 2,340 ft. **Prop:** 1 claim: Black Jack. **Owner:** Gold Bond Mining Co. (1951 ——). La Rica Consolidated Mining Co. (1906). Washington Meteor Mining Co. (1908-1911). Blewett Mine Leasing Co. (1910). **Ore:** Gold, mercury. **Ore min:** Free gold, native mercury, arsenopyrite, pyrite. **Deposit:** Quartz vein cutting serpentine. **Dev:** 1,300-ft. adit, a shorter adit, several winzes and raises. **Assays:** 3,000 tons of ore av. $10 Au. **Prod:** 3,000 tons of ore about 1900, $4,000 in 1940. **Ref: 63,** p. 73. **67,** p. 10. **88,** p. 60. **97,** 1906, p. 366. **129,** p. 275. **161,** pp. 84-89.

Blewett (130)

(See also Black Jack, Peshastin, Gold Bond.)

Loc: SE¼ sec. 2, (22-17E), Blewett dist. **Owner:** Gold Bond Mining Co., Spokane, Wash. (1952 ——). Amalgamated Gold Mines Co. (1918-1926). United States Mines & Metals Corp. (1932). Washington Gold Mines Corp. (1932). **Ore:** Gold. **Prod:** Amount of early shipments not known. Shipped 1939-1940. **Ref:** **97,** 1910, p. 603; 1931, p. 476; 1940, p. 476; 1941, p. 472. **98,** 1920-1926. **104,** 6/15/32, p. 27. **106,** 5/19/32; 8/18/32, p. 6.

Blinn (151)

(See also Gold Bond.)

Loc: NE¼ sec. 3, (22-17E). Claims extend from the head of Culver Gulch westward to and across Nigger Cr. **Access:** Road up Nigger Cr. **Prop:** 5 patented claims: Shafer, Olympia, Pole Pick No. 3, Seattle, Vancouver. **Owner:** Gold Bond Mining Co. (1936 ——). Cascade Mining Co. (1911). Bonanza Gold Mine Co. (1934). **Ore:** Gold. **Ore min:** Pyrite, chalcopyrite. **Gangue:** Quartz, calcite. **Deposit:** A 1- to 5-ft. quartz-calcite vein cuts serpentine and greenstone. **Dev:** 1 long adit and 1,000 ft. of shorter ones. **Prod:** In 1880 ore was milled in a 2-stamp Harrington mill. Produced 1936, 1937. **Ref: 67,** p. 10. **104,** 11/15/36, p. 27; 1/15/37, p. 28. **113,** 7/1/37, p. 16. **158. 161,** pp. 92-93.

Blue Bell (I. X. L.) (131)

Loc: SW¼ sec. 1, (22-17E), on E. side of Peshastin Cr. ¼ mi. N. of Blewett. **Access:** Road. **Owner:** Thad Neubauer, Blewett, Wash. (1934). **Ore:** Gold. **Ore min:** Pyrite, free gold. **Gangue:** Quartz, calcite. **Deposit:** No well-defined veins. Seams in serpentine are filled with vein and ore minerals. **Dev:** 450 ft. of adits. **Prod:** Some small pockets yielded high values. **Ref: 63,** p. 74. **67,** p. 10. **158. 161,** p. 99.

Bobtail (Wye) (132)

Loc: SW¼ sec. 2, (22-17E), in Culver Gulch, Blewett dist. **Elev:** 3,100 ft. **Access:** Road and trail. **Prop:** 1 claim. **Owner:** Washington Meteor Mining Co. (1908-1911). Blewett Mine Leasing Co. (1910). **Ore:** Gold. **Deposit:** Lenticular quartz vein in serpentine. Small ore body of medium grade. **Dev:** Two short adits, one 50 ft. long. **Prod:** Has produced. **Ref: 67,** p. 14. **129,** p. 272. **143,** pp. 78-79. **144,** p. 4. **161,** pp. 84-89.

Butte (116)

Loc: Secs. 25 and 26, (27-22E), Chelan Butte dist. **Owner:** Chelan Butte Gold Mining Co. (1909-1915). **Ore:** Gold. **Prod:**

Amount not known. **Ref: 67**, p. 20. **105**, 1898, p. 311; 1907, pp. 584, 807. **114**, no. 5, 1909, p. 84. **116**, no. 10, 1907, p. 18.

Cook-Galbraith (121)

Loc: T. 25 N., R. 20 E., about 3 mi. up Entiat R. from its mouth. Near top of ridge N. of Entiat R. **Elev:** 2,160 ft. **Access:** Road. **Prop:** 80 acres deeded land. **Ore:** Gold. **Ore min:** Free gold. **Gangue:** Quartz. **Deposit:** A 7- to 18-in. quartz vein in granodiorite. **Dev:** Open cuts. **Assays:** 2 assays gave values of $23.80 and $68.60 per ton (1935). **Ref: 67**, p. 25. **158**.

Culver (133)

Loc: SW¼ sec. 2, (22-17E), Blewett dist. **Elev:** 3,250 to 4,100 ft. **Access:** Road and trail. **Prop:** 1 claim. **Owner:** Washington Meteor Mining Co. (1911). Culver Gold Mining Co. (1891). Blewett Gold Mining Co. (1892-1894). Blewett Mining & Milling Co. (1895). Warrior General Co. (1896). Chelan Mining & Milling Co. (1902). Washington Meteor Mining Co. (1908-1909). Blewett Mining & Leasing Co. (1910). **Ore:** Gold, silver. **Ore min:** Free gold, arsenopyrite, pyrite. **Gangue:** Quartz, calcite, talc. **Deposit:** Ore occurs as lenses in a quartz-calcite vein enclosed in serpentine. **Dev:** 14 adits aggregating more than 4,500 ft. of length, and numerous raises and stopes. **Assays:** $5 to $500 Au. **Prod:** About $300,000 by 1902. Since then production has been listed with that for Washington Meteor Mining Co. **Ref: 12**, pp. 11-12. **63**, p. 72. **67**, p. 10. **88**, p. 58. **143**, pp. 78-79. **144**, p. 9. **161**, pp. 84-89.

Diamond Dick

(See Black and White.)

Eastern Star

(See Olden.)

Eureka (Golden Cherry, Golden Chariot) (134)

Loc: SW¼ sec. 1, (22-17E), on E. side of Peshastin Cr., 2,000 ft. S. of Blewett. **Elev:** 2,400 ft. **Prop:** 1 claim. **Owner:** Tip Top Mining Co. (1910). Phoenix Mining & Milling Co. (1902). **Ore:** Gold. **Deposit:** A 12- to 18-in. vein in serpentine. **Dev:** 970 ft. of underground work. **Assays:** Much of the ore assays as high as $30 Au. **Prod:** Amount not known. **Ref: 63**, p. 73. **67**, pp. 10-11. **88**, p. 57. **161**, p. 96.

Fraction (135)

Loc: SW¼ sec. 1, (22-17E), Blewett dist. **Ore:** Gold. **Prod:** Amount not known. **Ref. 67**, p. 11. **116**, no. 4, 1908, p. 74. **143**, p. 79. **144**, p. 9. **158**.

Galbraith

(See Cook-Galbraith.)

Gold Bond

(See also Black Jack, Blinn, La Rica, Olympia, Phipps, Pole Pick, Wilder.)

Loc: Blewett dist. **Prop:** 7 patented claims: Seattle, Vancouver, Shafer Extension No. 2, Olympia, Pole Pick No. 1, Pole Pick No. 2, Davidson; 17 unpatented claims. **Owner:** Gold Bond Mining Co., Spokane, Wash. (1937 ——). **Ore:** Gold. **Ore min:** Free gold. **Gangue:** Quartz, calcite. **Prod:** 1936-1937, 1941-1942, 1946-1951. **Ref: 97,** 1938, p. 458. **104,** 11/15/36, p. 27; 1/15/37, p. 28. **133,** p. 34.

Gold King

(See Golden King.)

Gold Knob (153)

Loc: Secs. 16 and 21, (22-20E), Squaw Saddle area, Wenatchee dist. **Access:** Road. **Prop:** All of sec. 16, approx. 20 acres of sec. 21. **Owner:** Leased from State by J. J. Keegan, Wenatchee, Wash. (1951 ——) and subleased by Anaconda Copper Mining Co. (1952). **Ore:** Gold, silver. **Deposit:** Gold- and silver-bearing quartz stockwork in Swauk sandstone. **Dev:** Several diamond drill holes, approx. 800 ft. of crosscutting through silicified zones, an approx. 90-ft. winze. **Assays:** Up to 1952 all assays have been far below ore grade. **Ref: 133,** p. 34. **150,** p. 32. **158.**

Golden Chariot

(See Eureka.)

Golden Cherry

(See Eureka.)

Golden Eagle (Lucky King) (136)

Loc: SE¼ sec. 2, (22-17E), on N. side of Culver Gulch ½ mi. above its junction with Peshastin Cr. **Prop:** 6 claims: Golden Eagle, Golden Eagle Fraction, Lone Jack, Lucky King, Leroy, Golden Eagle mill site. **Owner:** Golden Eagle Mining Co. (1911-1915). **Ore:** Gold. **Ore min:** Pyrite, free gold. **Gangue:** Quartz, calcite, talc. **Deposit:** A 1- to 3-ft. vein composed of quartz, calcite, and talc is impregnated with pyrite. **Dev:** Upper crosscut and drift total 175 ft. There is a raise to the surface. A lower crosscut totals 650 ft. **Assays:** 2 assays across the vein from upper adit near the raise gave $3.75 and $4.10 per ton. **Prod:** $2,000 reported. **Ref: 67,** p. 11. **88,** p. 59. **97,** 1910, p. 603; 1914, p. 647; 1915, p. 568. **105,** 1913, p. 909. **158. 161,** pp. 94-95.

Golden King (Wenatchee, Squillchuck, Gold King) (154)

Loc: Near center sec. 22, (22-20E), on W. side of Squillchuck Cr. **Elev:** 1,000 ft. **Access:** Good road to property. **Prop:** 2 claims. **Owner:** Lovitt Mining Co., Inc., Wenatchee, Wash. (1949 ——).

V. Carkeek (1885). Golden King Mining Co. (1894). Wenatchee Mining Co. (1910-1928). J. J. Keegan (1928-1949). American Smelting & Refining Co. (1938-1939). Knob Hill Mines, Inc. (1944-1946). **Ore:** Gold, silver. **Ore min:** Pyrite. **Gangue:** Quartz, calcite, siderite. **Deposit:** Quartz gash veins in a 200- to 800-ft. wide silicified zone, in sandstone of the Swauk formation. **Dev:** A quarry, 2 main adits, and a level from a 500-ft. inclined shaft total about 2 mi. of workings. **Improv:** Offices, dries, warehouses, compressor houses, machine shop, and equipment for 200-ton operation (1955). **Assays:** 3 years' production (1944-1946) av. $2.39 Au, Ag. 1950 shipments av. 0.74 oz. Au, 0.66 oz. Ag. 1949-1953 shipments av. 0.506 oz. Au, 0.59 oz. Ag. Highest monthly av. shipments ran 1.27 oz. Au. **Prod:** 240 tons yielded $1,600 in 1894; 170 tons 1910; 20,000 tons 1938-1939; 6,216 tons 1944-1946; 135,000 tons Aug. 1, 1949-Aug. 1, 1952. 5½ years' production through 1954 was valued at $4,432,884. A total of 57,689 tons of ore valued at $836,782 in 1953 yielded 23,136 oz. Au, 32,242 oz. Ag. **Ref: 23-A.** **63,** p. 71. **67,** p. 46. **105,** vol. 68, 1894, p. 382; vol. 101, 1910, p. 63. **158.**

Hidden Treasure (112)

Loc: NW¼SW¼ sec. 17, (30-21E), just W. of Chelan-Methow summit. **Prop:** 3 unpatented claims. **Owner:** J. H. Baker, Joe Baker, and Norman Lindsley, Chelan, Wash. (1946). **Ore:** Gold, silver. **Gangue:** Talc, chlorite. **Deposit:** Hydrothermally altered rock along contact of granitic rock with hornblendite. **Dev:** 20-ft. shaft. **Assays:** Surface samples assay $5 Au, Ag. **Ref: 51,** p. 6. **158.**

Holden (Howe Sound, Irene) (113)

Loc: Secs. 18 and 19, (31-17E) and secs. 12 and 13, (31-16E), on Railroad Cr. **Elev:** 3,435 ft. at haulage level. **Access:** Boat from Chelan to Lucerne and 12 mi. of good road from there to mine. **Prop:** 13 patented and 78 unpatented claims. **Owner:** Howe Sound Co., New York, N. Y. (1937 ——). J. H. Holden (1892-1896). Chelan Copper Co. (1907). Holden Gold & Copper Co. (1901-1924). Lake Chelan Copper Co. (1925-1932). Britannia Mining & Smelting Co., Ltd. (1928). Chelan Copper Mining Co. (1930-1936). **Ore:** Copper, gold, zinc, silver. **Ore min:** Chalcopyrite, pyrrhotite, pyrite, sphalerite, galena, magnetite, chalcocite, malachite, native copper, molybdenite, pitchblende (?), scheelite. **Gangue:** Silicified metamorphosed sediments. **Deposit:** Zone of sulfide disseminations 20 to 75 ft. wide has exposed length of 2,500 ft. and depth of 2,500 ft. Ore occurs in a roof pendant of metamorphic rocks cut by granitic dikes. Slight amount of radioactive mineralization in footwall zone in W. end of mine at 1,950- and 2,325-ft. levels. **Dev:** 247,566 ft. of drifts, crosscuts, and raises; 231,922 ft. of core drilling (1951). **Improv:** 2,000-ton flotation mill, modern camp for 450 men, roads, docks, tug, barges, and all necessary facilities. **Assays:** Mill feed av. 1.45% Cu, 0.09 oz. Au, 0.344 oz. Ag, 1.02% Zn in 1940, and 0.78% Cu, 0.44 oz. Au, 0.213 oz. Ag, 0.48% Zn in 1951. Samples from basic dike outside of ore zone assayed 0.2% to 0.46% Ni. Radiometric tests showed 0.019% U_3O_8 equivalent in one sample. **Prod:** 1938-1955. From 1938 to 1951 8,320,497 tons of ore were produced. 1950 production was 5,005 tons Cu, 2,531 tons Zn. 1951 production was 4,015 tons Cu, 1,958 tons Zn, 24,205 oz. Au, 117,437 oz. Ag from 550,530 tons of ore. **Ref: 1-A**, vol. 163, pp. 73-95. **37**, p. 16. **63**, p. 82. **67**, pp. 35-36. **88**, pp. 55-56. **93**, Ch. III, p. 13. **97**, 1929, 1930, 1937-1952. **98**, 1920-1926. **104**, 1/15/32, p. 29; 12/30/32, p. 24; 6/15/34, p. 29; 6/30/34, p. 23; 11/15/36, p. 27. **105**, 1907, p. 41. **108**, 11/39, p. 30; 5/40; 6/40. **113**, 6/17/37, p. 7. **114**, no. 5, 1909. **133**, p. 35. **148**. **158**. **159**, p. 137. **175**.

Howe Sound
(See Holden.)

Hummingbird (137)

Loc: SE¼ sec. 2, (22-17E), in Culver Gulch, Blewett dist. **Elev:** 2,900 ft. **Access:** Road and trail. **Prop:** 1 claim. **Owner:** Washington Meteor Mining Co. (1908-1911). **Ore:** Gold, copper, silver. **Ore min:** Free gold, arsenopyrite, pyrite, chalcopyrite. **Gangue:** Quartz, calcite, talc. **Deposit:** Quartz-calcite vein cutting serpentine. **Dev:** 650-ft. crosscut and drift with stopes. **Prod:** Amount not known. **Ref:** 13, p. 137. **67**, p. 11. **129**, p. 272. **143**, pp. 78-79. **144**, p. 9. **161**, pp. 84-89.

I. X. L.
(See Blue Bell.)

Irene
(See Holden.)

Ivanhoe
(See Wilder.)

Kingman and Pershall (117)

Loc: On Chelan Butte. **Ore:** Gold. **Ore min:** Free gold. **Deposit:** 4-ft. vein. **Assays:** One assay as high as $2,000 per ton. **Prod:** $15,000 reported in the 1890's. **Ref:** 67, p. 20. **158**.

La Rica (138)
(See also Black Jack, Peshastin, Gold Bond.)

Loc: Sec. 2, (22-17E), Blewett dist. **Prop:** 3 claims: Peshastin, Keynote, Black Jack. **Ore:** Gold. **Ref:** 97, 1916, p. 610; 1921, p. 462. **98**, 1925, p. 1801. **161**, p. 89.

Lucky King
(See Golden Eagle.)

Lucky Queen (150)

Loc: SW¼ sec. 1, (22-17E), about ¼ mi. N. of Blewett on E. side of Peshastin Cr. **Access:** On highway U. S. 97. **Prop:** 2 claims: Lucky Queen, Bee Queen. **Owner:** A. Neubaur, Blewett, Wash. (1938). Lucky Queen Mining Co. (1915). **Ore:** Gold, silver. **Ore min:** Free gold, chromite. **Gangue:** Talc, calcite, quartz. **Deposit:** Shear zone in serpentine from a stringer to 3 ft. in width. One small segregation of chromite encountered. **Dev:** 545-ft. adit, 412-ft. adit. **Prod:** About $1,000 by 1901. **Ref:** **67**, p. 11. **88**, p. 58. **116**, no. 4, 1908, p. 74. **143**, p. 79. **144**, p. 9. **161**, pp. 97-99.

North Star (139)

Loc: SE¼ sec. 2, (22-17E), on S. side and near upper end of Culver Gulch, Blewett dist. **Prop:** 1 claim. **Owner:** Golden Eagle Mining Co. (1911-1915). **Ore:** Gold. **Ore min:** Pyrite, arsenopyrite. **Gangue:** Quartz, calcite. **Deposit:** A 1- to 8-ft. fissure vein cutting serpentine. **Dev:** 125-ft. adit with stopes and raises, and 2 other adits totaling 320 ft. of length. **Assays:** Assays at various intervals across the vein av. $1 to $15 per ton. Av. of 1,000 tons shipped was $20 per ton. **Prod:** 1,000 tons ore prior to 1907. Produced 1915. **Ref:** **67**, p. 12. **97**, 1915, p. 568. **161**, pp. 93-94.

Olden (Eastern Star) (140)

Loc: SE¼ sec. 2, (22-17E), W. of the Black Jack and immediately S. of the Peshastin mine, Blewett dist. **Owner:** John Olden (1902). **Ore:** Gold. **Deposit:** 2 veins varying in width from 1 to 6 ft. **Dev:** 350 ft. of underground work. **Assays:** 500 tons av. about $5 Au. **Prod:** 500 tons of ore by 1902. **Ref:** **67**, p. 12. **88**, p. 59. **116**, no. 4, 1908, p. 74. **143**, p. 79. **144**, p. 9.

Olympia (141)

(See also Gold Bond.)

Loc: NW¼ sec. 2, (22-17E), about 600 ft. below and due W. of U. S. Mineral Monument at head of Culver Gulch, Blewett dist. **Prop:** 1 claim (part of Blinn property). **Owner:** Gold Bond Mining Co., Spokane, Wash. (1951——). **Ore:** Gold. **Ore min:** Pyrite, arsenopyrite, chalcopyrite. **Gangue:** Quartz, calcite. **Deposit:** Small quartz lenses in a sheer zone in serpentine. **Dev:** 910 ft. of crosscut and drifts. **Ref:** **63**, pp. 72-73. **67**, p. 17.

Ontario (152)

Loc: Sec. 4, (22-17E), N. of Nigger Cr. **Ore:** Gold, nickel, cobalt, copper. **Ore min:** Sulfides. **Deposit:** A wide mineralized zone in serpentine. **Dev:** 2 short adits and a shaft. **Assays:** $7 to $8 Au, 3% Ni, 3½% Cu. **Ref:** **43**, 1895, p. 184. **63**, p. 76. **67**, pp. 17-18, 29. **105**, 1895, p. 399. **141**, p. 63.

Pangborn (118)

Loc: SW¼ sec. 25, (26-20E), 1½ mi. E. of the Rex mine, Entiat dist. **Access:** Good road to mine. **Prop:** 80 acres. **Owner:** Leased from State by P. C. Pangborn, Wenatchee, Wash. (1938). **Ore:** Gold, silver. **Ore min:** Free gold. **Deposit:** Several quartz veins from 3 in. to 3 ft. wide cut decomposed gneiss. **Dev:** 180-ft. adit, 60-ft. shaft, 40-ft. shaft, numerous open cuts and short adits. **Assays:** Ore that was milled ran about ½ oz. Au. **Prod:** Amount not known. **Ref: 67,** p. 25. **158.**

Peshastin (Blewett, La Rica) (142)

(See also Blewett, La Rica, Gold Bond.)

Loc: SE¼ sec. 2, (22-17E), in lower Culver Gulch 1,200 ft. W. of old town of Blewett. **Elev:** 2,400 to 2,480 ft. **Access:** Road. **Prop:** 8 claims. **Owner:** Gold Bond Mining Co., Spokane, Wash. (1952———). Amalgamated Gold Mines Co. (1918-1925). La Rica Consolidated Mining Co. (1906). Washington Meteor Mining Co. (1908-1911). **Ore:** Gold. **Ore min:** Free gold, arsenopyrite, chalcopyrite, pyrite, galena, stibnite. **Gangue:** Quartz, calcite. **Deposit:** Lenticular lenses filling a shear zone in serpentine. Some lenses were 100 ft. long and as much as 8 ft. thick. **Dev:** 3 adits, the Meteor with 700 ft. of drifts, the Peshastin with 1,300 ft. of drifts and crosscuts, and the Draw with 480 ft. of workings. Many stopes in the adits. **Assays:** 22 tons shipped in 1940 returned 0.81 oz. Au, 0.13 oz. Ag, 0.09% Cu. **Prod:** About $60,000 by 1902, 22 tons in 1940. **Ref: 63,** p. 73. **67,** p. 12. **88,** p. 58. **97,** 1906, p. 366; 1921, p. 426. **98,** 1920-1926. **112,** 1918, p. 164. **116,** no. 4, 1908, p. 74. **129,** pp. 271-275. **143,** p. 79. **144,** p. 9. **158. 159,** p. 128. **161,** pp. 84-89.

Phipps (143)

(See also Gold Bond.)

Loc: SW¼ sec. 2, (22-17E), on Nigger Cr. drainage just over the ridge from head of Culver Gulch, Blewett dist. **Access:** Tractor road up Culver Gulch. **Owner:** Gold Bond Mining Co., Spokane, Wash. (1938 ———). **Ore:** Gold. **Ore min:** Free gold, chalcopyrite, arsenopyrite, pyrite. **Deposit:** Quartz vein varying from a thin seam to 5 ft. in width cuts serpentine. **Dev:** Several hundred feet of crosscuts, drifts, and stopes. **Prod:** Amount not known. **Ref: 67,** p. 12.

Phoenix (144)

Loc: Sec. 2, (22-17E), in Culver Gulch, Blewett dist. **Ore:** Gold. **Gangue:** Quartz. **Dev:** 3 drifts with stopes. **Assays:** Av. $20 Au on a large tonnage. **Prod:** 1,000 tons of ore prior to 1897. **Ref: 63,** p. 73. **67,** p. 13.

Pole Pick (145)

(See also Gold Bond.)

Loc: SW¼ sec. 2, (22-17E), on S. side of Culver Gulch about ½ mi. W. of Blewett. **Access:** Tractor road. **Prop:** 1 patented claim: Pole Pick No. 1. **Owner:** Gold Bond Mining Co. (1943 ——) leasing to Calton Mining Co. (1949-1952). Ellinor Mining Co. (1911). **Ore:** Gold. **Ore min:** Free gold, pyrite. **Gangue:** Quartz. **Deposit:** Main vein is 1 to 4 ft. wide. 2 subsidiary veins also occur. All 3 are in serpentine. **Dev:** 2 crosscuts with raises, drifts, and stopes total about 2,000 ft. of length. **Assays:** Ore extracted before 1901 is said to av. $10 to $132 Au. Ore mined 1942-1947 av. $35.82 Au. **Prod:** Est. production by 1901 was 8,000 tons of ore valued at $70,000. 1937, 1941, 1942 (291 tons), 1946 (90 tons), 1947 (81 tons), 1948 (19 tons), 1949 (20 tons), 1950-1951. **Ref:** 13, p. 137. **63,** p. 73. **67,** p. 13. **68,** p. 6. **88,** p. 59. **97,** 1935, p. 352. **133,** p. 31. **143,** p. 79. **144,** p. 9. **158. 161,** p. 92.

Pole Pick No. 2

(See Alta Vista.)

Prospect (146)

Loc: SW¼ sec. 2, (22-17E), on N. slope of ridge leading down to Culver Gulch, Blewett dist. **Prop:** 5 claims: Sunset, Sunset Extension, Redjacket, Katy, Lone Star. **Owner:** Prospect Mining & Milling Co. (1911). **Ore:** Gold, silver. **Gangue:** Quartz, calcite, talc. **Deposit:** Oxidized vein in serpentine. **Dev:** Small openings on all of the claims. **Prod:** Ore was treated in an arrastre in "early days." **Ref:** 67, p. 13. **161,** pp. 99-100.

Red Cap (114)

Loc: Sec. 9, (30-16E), on Phelps Ridge, Chiwawa dist. **Prop:** 20 claims. Probably part of the property held by the Royal Development Co. (1941). **Owner:** Una Mining & Milling Co., Seattle, Wash. (1897). **Ore:** Gold, silver, copper. **Ore min:** Chalcopyrite, pyrite, arsenopyrite. **Dev:** 52-ft. crosscut. **Assays:** $3.50 to $72 Au. **Ref:** 63, p. 78. **67,** p. 23.

Red Hill (115)

Loc: Sec. 15, (30-16E), on Phelps Ridge, Chiwawa dist. **Prop:** 10 claims. Probably part of the property held by the Royal Development Co. (1941). **Owner:** Red Hill Mining Co. (1897). **Ore:** Gold, silver, copper. **Ore min:** Copper and iron sulfides, arsenopyrite. **Dev:** 2 short adits. **Assays:** $2.50 to $29 Au, Ag. **Ref:** 63, p. 78. **67,** p. 23.

Rex (Rogers) (120)

Loc: N½ sec. 36, (26-20E), in Crum Canyon, tributary to Entiat R. **Access:** 12 mi. by road to railroad at Entiat. **Prop:** 80 acres. **Owner:** State land leased to Bert Rogers, Entiat, Wash.

(1943). Wenatchee Gold Mining Co. (1906-1922). **Ore:** Gold, silver. **Ore min:** Pyrite. **Deposit:** 2 oxidized quartz veins 3 to 12 in. wide in decomposed gneiss. **Dev:** 3 adits totaling about 500 ft. of length. **Improv:** A 2-stamp mill (1938). **Assays:** Most of the ore milled av. about 1 oz. Au. **Prod:** More than $170,000 by 1930. Small amounts in 1933, 1934, and 1940. **Ref: 43,** 3/13, p. 592. **67,** p. 25. **97,** 1907-1908, 1911-1915, 1917, 1934, 1935, 1940. **98,** 1920-1925. **104,** 6/15/32, p. 27. **105,** 1906, p. 115. **106,** 5/19/32. **112,** p. 211. **158.**

Rogers

(See Rex.)

Sandell (147)

Loc: SE¼ sec. 2, (22-17E), in Culver Gulch, Blewett dist. **Elev:** 2,660 ft. **Prop:** 1 claim, part of Washington Meteor Holdings. **Owner:** Washington Meteor Mining Co. (1908-1911). Blewett Mine Leasing Co. (1910). **Ore:** Gold. **Ore min:** Free gold, arsenopyrite, pyrite. **Gangue:** Quartz, calcite, talc. **Deposit:** Quartz-calcite vein cutting serpentine. 1 ore shoot mined av. 2 to 6 ft. thick. **Dev:** 300 ft. of crosscut and a large amount of drifting on the vein. **Prod:** Amount not known. **Ref: 67,** p. 13. **129,** p. 274. **161,** pp. 84-89.

Squillchuck

(See Golden King.)

Sunshine (119)

Loc: Sec. 36, (26-20E), Entiat dist. **Access:** 12 mi. by road to railroad at Entiat. **Prop:** 40-acre State lease. **Owner:** Leased by P. C. Pangborn, Wenatchee, Wash. (1935). **Ore:** Gold. **Ore min:** Free gold. **Dev:** 100-ft. shaft, 730-ft. adit. **Assays:** $14.96 to $306.00 Au. **Ref: 158.**

Tip Top (148)

Loc: SE¼ sec. 1, (22-17E), E. of the Eureka mine, Blewett dist. **Prop:** 1 claim. **Owner:** Tip Top Mining Co. (1911). **Ore:** Gold. **Ore min:** Free gold. **Gangue:** Quartz. **Deposit:** Vein av. 2½ ft. in width cuts serpentine and brecciated rock of the Hawkins formation. **Dev:** 500 ft. to 600 ft. of adit. Ore treated in an arrastre in 1901. **Assays:** Oxidized ore near surface av. $40 per ton; deeper ore av. $25 per ton. **Prod:** About $10,000 by 1901, small amount in 1940. **Ref: 63,** p. 74. **67,** p. 13. **88,** pp. 57-58. **97,** 1941, p. 472. **116,** no. 4, 1908, p. 74. **143,** p. 79. **144,** p. 9. **161,** pp. 95-96.

Wenatchee

(See Golden King.)

White Elephant
(See Wilder.)

Wilder (Ivanhoe, White Elephant) (149)
(See also Gold Bond.)

Loc: SW¼ sec. 2, (22-17E), on N. side of Culver Gulch and S. side of Nigger Cr., Blewett dist. **Prop:** 3 claims: Kennilworth, Amber Glee, Ivanhoe. **Owner:** Gold Bond Mining Co., Spokane, Wash. (1937——). **Ore:** Gold, silver, copper. **Deposit:** A 2- to 6-ft. vein. **Dev:** Several crosscuts and drifts total about 500 ft. of length. **Assays:** Tr. to $72 Au, with a very little Ag and Cu. Av. assay is $4.50 per ton. **Prod:** Est. 150 tons of ore prior to 1911. 1915, 1917, 1937. **Ref:** 63, p. 73. 67, p. 14. 88, p. 59. 97, 1915, p. 568; 1916, p. 610; 1917, p. 500. 104, 11/15/36, p. 27; 1/15/37, p. 28. 106, 6/18/31, p. 19. 113, 7/1/37, p. 16. 158. 161, pp. 96-97.

Wye
(See Bobtail.)

CLALLAM COUNTY

Port Angeles (1)
Loc: On Melick farm, between Ennis Cr. and head of White Cr., probably in sec. 23, (30-6W). **Access:** Road. **Prop:** 43 acres of deeded farm land. **Owner:** Mrs. Grace Melick, Port Angeles, Wash. (1950). **Ore:** Gold and silver reported. **Ore min:** Pyrite. **Gangue:** Quartz. **Assays:** One assay showed about $1 Ag, $123 Au, but two other assays showed only $1.50 and $2.10 Au. **Ref:** 158.

CLARK COUNTY

Golden Wonder (82)
Loc: NE¼NE¼ sec. 32, (6-4E), 2 mi. E. of Yale bridge on Lewis R. **Elev:** 300 ft. **Access:** Aerial tram across Lewis R. from Frazier ranch. **Prop:** 120 acres of deeded land. **Owner:** V. V. Rand, Vancouver, Wash. (1942). Golden Wonder Mining Co. (1934). **Ore:** Gold, mercury. **Ore min:** Pyrite, cinnabar. **Gangue:** Calcite, gouge, quartz. **Deposit:** Weakly mineralized shear zone in altered tuff. Zone is 150 to 200 ft. wide at one place. Cinnabar occurs as a few scattered crystals. **Dev:** Adit caved at 600 ft. from portal, and other short adits and open cuts also caved. **Assays:** Said to assay 0.02 oz. Au. **Ref:** 158.

Silver Star (83)
Loc: Secs. 14, 15, 22, and 23, (3-4E), on W. slope of Silver Star Mtn. **Elev:** 1,500 ft. **Access:** 9 mi. by road to railroad at Yacolt. **Prop:** 1,040 acres. **Owner:** Silver Star Mining Co., R. Demott, Portland, Oreg. (1930 ——). **Ore:** Copper, gold, silver, lead, zinc,

nickel. **Ore min:** Chalcopyrite, pyrite, siderite, sphalerite. galena, magnetite. **Gangue:** Quartz and altered country rock. **Deposit:** Mineralized extrusive rock. About 2 tons of ore on the dump. **Dev:** 125-ft. adit, 227-ft. adit. **Assays:** Assays show 0.08 to 0.24 oz. Au, 0.50 to 2.8 oz. Ag, 2.32% Cu, 3.6% to 12.8% Zn, and 0.34% to 0.4% Pb. A 3.8-ft. channel sample showed 0.9% Zn. 0.3% Cu, 0.9 oz. Ag. **Ref:** 111, p. 10. **158.**

COWLITZ COUNTY

Green Mountain (81)

Loc: Sec. 6, (5-1E), near Woodland. **Prop:** Deeded land. **Owner:** Margaret Hurtienne, Kelso, Wash. **Ore:** Gold, silver, lead, zinc, copper. **Ore min:** Pyrite. **Deposit:** Fault in andesite, with weak hydrothermal alteration along the walls. Gouge and altered walls contain finely crystalline pyrite. **Dev:** Small open cut. **Assays:** 2 samples showed only traces of gold and very low values in silver, lead, zinc, copper. **Ref:** 157. **158.**

FERRY COUNTY

Advance (224)

Loc: Secs. 12 and 13, (36-32E), just S. of Old Republic mine, Republic dist. **Prop:** Patented. Part of Consolidated Mines and Smelting Co., Ltd. property. **Owner:** Consolidated Mines and Smelting Co., Ltd., Wenatchee, Wash. (1940). **Ore:** Gold, silver. **Ref:** 57, p. 2. **158.**

Anecia (205)

Loc: Sec. 33, (37-32E), Republic dist. **Owner:** Dr. May, Republic, Wash. (1940). **Ore:** Gold, silver. **Prod:** Unknown amount in 1940 and 1941. **Ref:** 97, 1940, p. 477; 1941, p. 473.

Apollo

(See California.)

Aurum

(See Day Mines.)

Ben Hur (206)

(See also Day Mines.)

Loc: Center of line between secs. 34 and 27, (37-32E). **Elev:** 2,800 ft. **Access:** About 2 mi. NW. of Republic by road on W. side of Eureka Gulch. **Prop:** 1 claim adjoining the Trade Dollar on the south. **Owner:** Day Mines, Inc. (1951 ——). Ben Hur Gold Mining Co. (1907-1908). Ben Hur Mining & Milling Co. (1908). Ben Hur Leasing Co. (1909-1924). Aurum Mining Co. (1935-1950). **Ore:** Gold, silver. **Ore min:** Pyrite. **Gangue:** Quartz, calcite. **Deposit:** 4-ft. quartz vein in propylitic latite porphyry. Composed of fine-grained banded quartz and 10% to 30% calcite. Vein said to extend the length of the claim. **Dev:** 300-ft. shaft, winze, and drifts on 3 levels. **Assays:** Smelter assays of ore shipped show $15 Au, 4.5 oz. Ag. In 1910, $6 to $10 per ton. **Prod:** $65,000 up to 1910, 1909-1915, 1918, 1933, 1949, 1950. **Ref: 7,** p. 163. **88,** pp. 20-21. **97,** 1910-1919, 1932-1934. **98,** 1918, 1920, 1925. **104,** 1/30/33, p. 19. **106,** 2/18/32. **114,** no. 5, 1909. **153,** pp. 50-52. **158.**

Black Tail (Hope) (207)

(See also Day Mines.)

Loc: Near E¼ cor. sec. 34, (37-32E), S. of the Lone Pine and N. of the Quilp. **Access:** 1 mi. NW. of Republic by road. **Prop:** 1 claim. **Owner:** Day Mines, Inc. (1951 ——). Hope Mining Co. (1910-1915). The Hope Co. (1918). Northport Smelting & Refining Co. (1920-1926). Surprise Mining Co. (1924-1926). **Ore:** Gold, silver. **Deposit:** Several quartz veins 2 to 6 ft. wide in quartz latite porphyry and propylitic andesite. **Dev:** Crosscut adit 300 ft. long, drifts, and 600-ft. inclined shaft. Workings total about 2,000 ft. **Assays:** 300 tons produced prior to 1902 ranged from $13 to $20 per ton. **Prod:** 1909-1910, 1912-1920, 300 tons prior to 1902. **Ref: 7,** p. 159. **88,** p. 20. **97,** 1910-1921, 1935. **98,** 1922-1926. **112,** p. 206. **114,** no. 5, 1909. **129,** pp. 186-187. **153,** pp. 56-57.

Blaine Republic

(See Republic.)

Boston and Butte

(See Butte and Boston.)

Butte and Boston (Boston and Butte) (225)

Loc: SW¼ sec. 12, (36-32E), Republic dist. **Owner:** Blaine-Republic Co. (1932). Anaconda Gold Mining & Reduction Co. (1914). Alliance Mining Co. (1918-1926). **Ore:** Gold, silver. **Deposit:** 2½- to 5-ft. quartz vein cuts andesite porphyry. Same vein as on Princess Maude property. **Dev:** 265-ft. shaft, 75-ft. shaft, 285-ft. adit, 35-ft. adit, and 400 ft. of drifts (1902). **Assays:** Av. $16 per ton, $14 of which is in gold. **Ref: 88,** p. 22.

California (Apollo) (229)

Loc: SW¼SW¼ sec. 20, (36-34E), on Wabash Cr., Republic dist. **Elev:** 4,200 ft. **Access:** 11 mi. E. of Republic by road. **Prop:** 2 claims: Colorado, Shonneen. **Owner:** New California Mining Co., Wenatchee, Wash. (1949). Apollo Consolidated Gold Mining Co. (1903-1925). **Ore:** Gold, silver. **Ore min:** Galena, chalcopyrite, sphalerite, malachite, azurite. **Deposit:** Quartz vein along fracture zone in greenstone and argillite. **Dev:** 525-ft. shaft, 80-ft. adit. **Assays:** $60 Au, Ag. **Prod:** 1901-1902, 1908, 1914-1916, 1927-1929, 1938-1939. **Ref: 7,** p. 201. **33,** 1907, p. 316. **68,** p.

14. **97**, 1908, 1915-1917, 1928-1930, 1939, 1940. **98**, 1922, p. 1633; 1925, p. 1803. **100**, 1901, pp. 18, 50, 88; 1902, pp. 14, 132; 1903, pp. 14, 35. **112**, 1918, p. 166. **158**.

Day Mines (Aurum)

(See also Surprise, Little Cove, Last Chance, San Poil, Ben Hur, Insurgent, Lone Pine, Tom Thumb, Trade Dollar, Black Tail, Pearl, Quilp, South Penn.)

Loc: Republic dist. **Prop:** At least 92 patented claims. **Owner:** Day Mines, Inc., Wallace, Idaho (1951 ——). Formerly operated through a subsidiary: Aurum Mining Co. (1935-1950). **Ore:** Gold, silver. **Ore min:** Tellurides, free gold. **Gangue:** Quartz, calcite. **Deposit:** Veins in andesite porphyry. **Dev:** Deepest workings 900 ft. (1939). **Assays:** Av. 0.15 oz. Au, 1 oz. Ag. **Prod:** Considerable. See individual mines for details. **Ref: 107**, 11/39.

El Caliph (208)

Loc: SW¼ sec. 34, (37-32E), just E. of the Morning Glory property, Republic dist. **Access:** 2 mi. by road to railroad. **Prop:** 1 patented claim: El Caliph. **Owner:** E. D. Hougland, Republic, Wash. (1941). **Ore:** Gold, silver. **Ore min:** Free gold, pyrite. **Gangue:** Quartz, calcite. **Deposit:** ½- to 18-in. vein cutting quartz latite and shales. Vein displaced by minor faults. **Dev:** Adit and shaft totaling 450 ft. (1902). **Assays:** 85 tons of ore produced prior to 1902 yielded $9,000. **Prod:** Est. $15,000 to $20,-000 to end of 1936. Produced 1916, 1933, 1934, 1937-1939. **Ref: 88**, p. 21. **97**, 1916, 1934, 1935, 1938-1940. **153**, p. 65. **158**.

Eureka

(See Quilp.)

Faithful Surprise

(See Morning Star.)

Flag Hill (226)

Loc: Secs. 1 and 2, (36-32E). **Access:** 1½ mi. from Republic by road. **Prop:** 8 unpatented claims. **Owner:** Flag Hill Mines, Inc. (1949 ——). **Ore:** Gold, silver, selenium. **Gangue:** Quartz, calcite. **Deposit:** Said to be a 5-ft. vein with an est. length of 1,500 ft. **Dev:** 1,500 ft. of adit, three 50-ft. shafts, and a 100-ft. shaft. Total length of workings about 4,000 ft. **Assays:** Typical assay said to be 0.21 oz. Au, 0.61 oz. Ag. **Prod:** Said to be 400 tons prior to 1940. 1941. **Ref: 28**, pp. 62-66. **69**, p. 7. **97**, 1939, p. 490; 1940, p. 477. **113**, no. 13, 1937, p. 7. **158**.

Golden Harvest (230)

Loc: Sec. 36, (36-32E). **Access:** 7 mi. from Republic by road. **Prop:** 7 patented claims. **Owner:** Otto Harrison, Wenatchee, Wash. (1941). Golden Harvest Mining Co. (1932-1937). **Ore:**

Gold, silver. **Deposit:** Said to be a 2½-ft. vein with an est. length of 150 ft. **Dev:** 1,600-ft. adit and several shafts. **Assays:** Typical assay said to be 0.21 oz. Au, 2.0 oz. Ag. **Prod:** 800 tons in 1937-1938. **Ref: 28,** pp. 112-116. **97,** 1939, p. 490. **104,** 6/15/35, p. 27; 2/15/35, p. 30; 9/30/36, p. 29; 1/30/37, p. 30. **106,** 1/21/32. **158.**

Golden Valley

(See Valley.)

Hawkeye (201)

Loc: NE¼ sec. 6, (37-34E), Belcher dist. **Prop:** 8 claims: Hawkeye, Hawkeye No. 3, Struck Luck, Eagle, Governor Rogers, Laughing Water, St. Bernard, Waterloo. **Owner:** Winnipeg Mining Co. (1907-1918). **Ore:** Gold, silver, copper, iron. **Prod:** 1907. **Ref: 33,** 1907, p. 1164; 1908, p. 1431. **98,** 1918, p. 148. **116,** no. 11, 1907, p. 27.

Hope

(See Black Tail.)

Ida May (209)

Loc: SW¼ sec. 34, (37-32E), Republic dist. **Access:** 2 mi. by road to railroad. **Prop:** 1 patented claim: Ida May. **Owner:** E. D. Hougland, Republic, Wash. (1941). **Ore:** Gold, silver. **Deposit:** Small vein. **Dev:** 50-ft adit. **Prod:** 1914. **Ref: 97,** 1914, p. 649. **158.**

Imperator

(See Quilp.)

Insurgent (221)

(See also Day Mines.)

Loc: Near W. line NW¼ sec. 35, (37-32E). **Access:** 1 mi. NW. of Republic by road. **Prop:** 1 fractional claim adjoining the Lone Pine on the east. **Owner:** Day Mines, Inc., Wallace, Idaho (1951 ——). Insurgent Gold Mining Co. (1908-1924). Aurum Mining Co. (1935-1950). **Ore:** Gold, silver. **Ore min:** Pyrite, gold. **Gangue:** Quartz. **Deposit:** Vein believed to be an offshoot of the Lone Pine vein cuts propylitic andesite. Ore shoot 30 ft. long and 2½ ft. wide now stoped out. **Dev:** Opened by adits on Lone Pine claim. **Prod:** 1908-1912, 1927. **Ref: 7,** p. 162. **97,** 1908-1912, 1927. **98,** 1920-1925. **104,** 10/30/35, p. 25; 2/29/36, p. 27. **112,** p. 184.

Iron Mask (210)

Loc: Sec. 32, (37-32E). **Access:** 1 mi. from Republic by road. **Prop:** 1 patented claim. **Owner:** P. B. Chapman, Republic, Wash. (1941). **Ore:** Gold, silver. **Deposit:** Said to be an 8-ft. vein with an est. length of 400 ft. **Dev:** 49-ft. adit and other workings total 150 ft. **Assays:** Typical assay said to be 0.12 oz. Au, 0.12 oz. Ag. **Ref: 28,** pp. 92-96.

Knob Hill (211)

(See also Mountain Lion, Rebate.)

Loc: W½SE¼ sec. 27, (37-32E), at head of Eureka Gulch. **Elev:** 2,700 ft. **Access:** About 2½ mi. by road from railroad at Republic. **Prop:** 13 patented claims, including Knob Hill, Mud Lake, Alpine, Lone Hand, Rebate, Mountain Lion. **Owner:** Knob Hill Mines, Inc., San Francisco, Calif. (1936 ——). Knob Hill Mining Co. (1910-1924). Balaklala Consolidated Copper Co. (1925). Balaklala Central Mining & Smelting Co. (1935). Mountain City Copper Co. (1935-1936). **Ore:** Gold, silver. **Ore min:** Free gold, tellurides, pyrite, arsenopyrite, stibnite, realgar, marcasite, tetrahedrite, polybasite, pyrargyrite, argentite. **Gangue:** Quartz, chalcedony, adularia, sericite, calcite, barite, graphite. **Deposit:** 4 parallel veins and a cross vein with mining widths of 5 to 15 ft. (1952). **Dev:** A 2-hoisting-compartment inclined 1,200-ft. shaft with 9 levels at 110-ft. vertical intervals. Abandoned open pits. **Improv:** 400-ton flotation-cyanidation mill, camp and office buildings 1955). **Assays:** 7,192 tons of ore av. 1.5 oz. Au, 4.5 oz. Ag. **Prod:** More than $10,000,000 by end of 1951. Produced 1937-1955. **Ref:** **7**, p. 164. **28**, pp. 57-61. **31**. **43**, 11/53. pp. 96-99. **63**, p. 109. **97**, 1910-1925, 1931, 1934-1949. **98**, 1918-1925. **104**, 8/15/35, p. 26; 10/30/35, p. 25; 1/30/36, p. 21; 11/30/36, p. 29. **107**, 12/39, p. 10; 1/40; 8/41, pp. 29-31. **112**, p. 187. **113**, 10/15/36, p. 11. **129**, pp. 175-180. **133**, p. 36. **141**, pp. 22, 34. **158**. **159**, p. 130. **173**, pp. 51-63. **174**, pp. 264-282.

Lame Foot

(See Valley.)

Last Chance (222)

(See also Day Mines.)

Loc: W½NW¼ sec. 35, (37-32E), on E. side of Eureka Gulch. **Access:** 1 mi. NW. of Republic by road. **Prop:** 1 claim: Last Chance. **Owner:** Day Mines, Inc., Wallace, Idaho (1951 ——). Lone Pine Surprise Consolidated Mining Co. (1918-1935). Aurum Mining Co. (1950). **Ore:** Gold, silver. **Ore min:** Free gold, tetrahedrite, pyrite. **Gangue:** Quartz, calcite. **Deposit:** Vein in andesite flow breccia av. 8 ft. in width. Vein filling consists of chalcedonic banded quartz, calcite, and fragments of country rock. Ore is largely removed above the 500-ft. level. **Dev:** 2-compartment vertical shaft 690 ft. deep, from which 1,200 ft. of drifts and crosscuts have been driven. Also extensive stoping. **Assays:** 24,000 tons of ore av. about $13.00 per ton. Ratio of precious metals av. 7.85 oz. Ag to 1 oz. Au. **Prod:** $3,000,000 by the end of 1923. 1940. **Ref:** **97**, 1914, 1915, 1918, 1919, 1922-1924, 1928, 1931, 1934, 1935. **98**, 1918-1926. **104**, 10/30/35, p. 25. **112**, p. 189. **129**, pp. 182-184. **158**.

Little Cove (212)

(See also Day Mines.)

Loc: NE¼ sec. 34, (37-32E). Adjoins the Pearl on the N., Republic dist. **Access:** Road. **Prop:** 1 claim. **Owner:** Day Mines, Inc., Wallace, Idaho (1951 ——). Little Cove Mining Co. (1918-1924). Aurum Mining Co. (1950). **Ore:** Gold, silver. **Gangue:** Chalcedonic quartz, calcite. **Prod:** 2 carloads valued at $1,450 prior to 1934. Produced 1934, 1939-1940. **Ref: 58,** p. 39. **97,** 1935, p. 352. **98,** 1922, p. 1653; 1925, p. 1822. **112,** p. 189. **158.**

Lone Pine (213)

(See also Day Mines.)

Loc: E½NE¼ sec. 34, (37-32E), on E. side of Eureka Gulch. **Access:** 1 mi. NW. of Republic by road. **Prop:** 1 claim: Lone Pine. **Owner:** Day Mines, Inc., Wallace, Idaho (1951 ——). Syndicated Deep Mines (1907-1908). Western Union Mines Co. (1914-1915). West Virginia Mining Co. (1915-1918). Republic Consolidated Mines Corp. (1915-1922). Northport Smelting & Refining Co. (1916-1926). Surprise Mining Co. (1924-1926). Aurum Mining Co. (1935-1950). **Ore:** Gold, silver. **Gangue:** Quartz, calcite. **Deposit:** 5 veins in propylitic andesite are from 2 to 14 ft. wide and consist of chalcedonic quartz traversed by narrow black crenulated ribbons. Most of the ore above the 500-ft. level stoped out. **Dev:** More than 2,500 linear ft. of underground workings, principally on 2 adits and a shaft. **Assays:** Ore av. $10 to $15 per ton. **Prod:** $137,000 to 1910, 1935. **Ref: 7,** pp. 160-162. **33,** 1908, p. 1294. **63,** p. 109. **77,** vol. 19, p. 93. **88,** pp. 19-20. **97,** 1910-1922, 1935. **98,** 1918-1926. **104,** 10/30/35, p. 25; 2/29/36, p. 27. **112,** p. 211. **116,** no. 11, 1907, p. 27; no. 12, 1907, p. 15; no. 5, 1908, p. 117; no. 6, 1908, p. 139. **129,** pp. 185, 187. **141,** p. 22. **153,** pp. 59-63. **158.**

Lucile Dreyfus

(See Morning Star.)

Mineral Hill

(See Morning Star.)

Morning Glory (Old Gold) (214)

Loc: S½SW¼ sec. 34, (37-32E), on Flag Hill. **Elev:** 3,000 ft. **Access:** ½ mi. by road from railroad near Republic. **Prop:** 1 patented claim and a fraction. **Owner:** C. E. Grove, R. R. Pence, R. M. Skidmore, Spokane, Wash. (1941). Morning Glory Mining Co. (1910-1915). **Ore:** Gold, silver. **Ore min:** Gold, pyrite, tellurides. **Gangue:** Quartz, calcite. **Deposit:** Vein of drusy banded quartz from a few inches to 2 or 3 ft. wide in quartz latite porphyry. Several rich pay shoots have been mined. **Dev:** Adit and air shaft with drifts said to aggregate 1,700 ft. **Assays:** Ore shipped carried values up to $400 Au, Ag, mostly Au. Tellurium is reported. **Prod:** $100,000 to 1936, 1937-1939. **Ref: 7,** p. 165. **28,** pp. 97-101. **88,** p. 21. **97.** 1914-1916, 1932-1935, 1938-1940. **153,** p. 52. **158.**

Morning Star (Lucile Dreyfus, Faithful Surprise, Mineral Hill, Virginia) (197)

(See also Surprise.)

Loc: SW¼ sec. 16, (40-34E). **Elev:** 2,300 ft. **Access:** 3 mi. by road S. of Danville on E. side of Kettle River. Railroad crosses property. **Prop:** 10 claims, 9 of which are patented: Mondamin, Tycoon, Minnehaha, Morning Star, Copper Bullion, Old Virginia, Alabama, Alabama Fraction, Copper Lady. **Owner:** Morning Star Mining Co. (1936 ——). Lucile Dreyfus Mining Co. (1908-1916). Mineral Hill Tunnel & Copper Mining Co. (1908-1918). Kettle River Mining & Lumber Co. (1918). Virginia Mining Co. (1918-1922). Chatterboy Mining Co. (1920-1926). **Ore:** Gold, silver, copper. **Ore min:** Free gold, pyrite, scheelite, chalcopyrite, pyrrohotite. **Deposit:** Quartz vein cutting serpentine is said to av. 2 ft. in width. **Dev:** 3,000-ft. haulage adit, 2 shafts of 220 and 325 ft. Total workings 12,000 ft. **Assays:** Av. 0.66 oz. Au, 1 oz. Ag, 2% Cu. **Prod:** $15,000 to 1910, $27,000 in 1917, about $15,000 in 1935. Produced 1936-1939, 790 tons 1940-1943. **Ref: 7,** p. 200. **28,** pp. 82-86. **33,** 1908, p. 892. **37,** pp. 25-26. **97,** 1913, 1916-1919, 1921, 1930, 1937-1941. **98,** 1920-1926. **104,** 10/30/36, p. 32. **112,** pp. 186, 191, 208. **113,** 10/15/36, p. 11; 2/18/37, p. 8. **129,** p. 202. **158.**

Mountain Lion (215)

(See also Knob Hill.)

Loc: W½ sec. 27, (37-32E), on E. side of the N. Fk. of Granite Cr. **Elev:** 3,000 ft. **Access:** 3 mi. NW. of Republic by road. **Prop:** 2 claims: Mountain Lion, Willamette. **Owner:** Knob Hill Mines, Inc., Republic, Wash. (1938 ——). Mountain Lion Gold Mining Co. (1898-1915). Mountain Lion Consolidated Co. (1934-1938). **Ore:** Gold, silver. **Ore min:** Gold, pyrite. **Deposit:** 3 parallel veins of banded quartz in andesite flow breccia. Productive vein is 10 to 12 ft. wide. **Dev:** 1,260-ft. adit and 700-ft. vertical shaft. Open pit mining methods began in 1941. **Assays:** Smelter assays of ore in 1904 and 1905 show $5 Au, 2 oz. Ag. **Prod:** $200,000 to 1910, 1936-1938. Production after 1938 is included with Knob Hill mine. **Ref: 7,** pp. 164-165. **43,** vol. 68, pp. 725-726. **88,** p. 18. **97,** 1916, 1932-1949. **104,** 1/30/33, p. 19; 4/15/34; 11/30/34, p. 23; 1/30/37, p. 28. **106,** 9/15/32. **113,** 10/15/36, p. 11. **133,** p. 36. **153,** pp. 63-64. **158. 159,** p. 130.

Old Gold

(See Morning Glory.)

Panama (199)

Loc: Near N. ¼ cor. sec. 6, (39-34E), Curlew area. **Owner:** Boston-New York Mines Co. (1924-1926). Phoenix Gold & Copper Mining & Milling Co. (1911-1918). National Lead-Silver Co. (1918-1922). **Ore:** Gold, silver, copper, lead, zinc. **Assays:** 22 samples showed nil to 4.6% Cu, 0.25 to 7.12 oz. Au, 1.22 to 89.7 oz. Ag. The av. was 1.71% Cu, 0.98 oz. Au, 24.80 oz. Ag. **Ref: 98,** 1918-1925. **116,** no. 2, 1911, pp. 40-41.

Pearl (216)

(See also Day Mines.)

Loc: Near center NE¼ sec. 34, (37-32E), adjoining Surprise claim on the N. **Access:** 1 mi. NW. of Republic by road. **Prop:** 1 claim. **Owner:** Day Mines, Inc., Wallace, Idaho (1951 ——). Syndicated Deep Mines (1907-1908). Pearl Consolidated Mining Co. (1909-1918). Republic Mines Corp. (1910-1912). Western Union Mines Co. (1914-1915). West Virginia Mining Co. (1915-1918). Republic Consolidated Mines Corp. (1915-1922). Northport Smelting & Refining Co. (1916-1926). Surprise Mining Co. (1924-1926). Aurum Mining Co. (1949-1951). **Ore:** Gold, silver. **Gangue:** Quartz. **Deposit:** 12-ft. vein (Surprise) continues through the claim its entire length of 1,500 ft. Wall rock is propylitic quartz latite porphyry. **Dev:** Diamond drilling. **Assays:** Said to assay $4.60 per ton. **Prod:** 1909-1922. **Ref: 7,** p. 159. **33,** 1908, p. 1294. **97,** 1909-1922. **98,** 1918-1926. **112,** p. 211. **113,** 12/15/36, p. 11. **116,** no. 11, 1907, p. 27; no. 12, 1907, p. 15; no. 5, 1908, p. 117; no. 6, 1908, p. 139. **129,** pp. 185, 188. **158.**

Princess Maude (Southern Republic) (227)

Loc: Near center sec. 12, (36-32E), near base of the E. slope of Copper Mtn. **Access:** 1 mi. SW. of Republic by road. **Prop:** 1 claim. **Owner:** Blaine-Republic Co. (1932). Southern Republic Mining Co. (1910). Anaconda Gold Mining & Reduction Co. (1914). Alliance Mining Co. (1915-1926). **Ore:** Gold, silver. **Ore min:** Pyrite. **Gangue:** Quartz, calcite, laumontite. **Deposit:** 2- to 4-ft. vein in propylitic andesite. Vein consists of vitreous white quartz showing lines of crustification parallel to the walls. **Dev:** 70-ft. incline and 200-ft. adit from which a 200-ft. and a 300-ft. drift have been driven. **Assays:** Ore shipped said to average $25 per ton. **Prod:** Has produced. **Ref: 7,** pp. 153-154. **88,** pp. 21-22. **98,** 1918-1926. **106,** 7/21/32, p. 7. **112,** pp. 164, 166. **114,** no. 5, 1909. **153,** pp. 52-53.

Quilp (Imperator, Eureka) (223)

(See also Day Mines.)

Loc: Near W. line SW¼ sec. 35, (37-32E), on E. side of Eureka Gulch. **Elev:** 2,800 ft. **Access:** 1 mi. NW. of Republic by road.

Prop: 1 patented claim: Quilp, originally located as the San Poil. **Owner:** Day Mines, Inc., Wallace, Idaho (1951 ——). Imperator Quilp Mining Co. (1909-1912). Quilp Gold Mining Co. (1915-1941). Eureka Mining & Milling Co. (1936). Aurum Mining Co. (1943-1951). **Ore:** Gold, silver. **Ore min:** Gold, pyrite, chalcopyrite, native silver. **Deposit:** Vein of chalcedonic banded quartz 7 to 8 ft. wide cuts propylitic andesite. **Dev:** 400-ft. shaft with several drifts and stopes. Winze sunk from 400-ft. to 500-ft. level. **Assays:** 20,000 tons of ore in 1906 had av. assay of 0.4 oz. Au, 5 oz. Ag. Av. assay on shipments in 1909 was 0.5 oz. Au, 4 oz. Ag. **Prod:** Total $720,938.70 to end of 1920; 1936; 22,402 tons 1937; 9,828 tons 1938; 1939-1940. **Ref: 7,** pp. 157-158. **28,** pp. 107-111. **88,** p. 19. **97,** 1909-1914, 1918-1920, 1922-1924, 1927, 1937-1940. **98,** 1918-1926. **104,** 6/30/36, p. 22; 11/20/36, p. 29; 12/15/36, pp. 26-27. **112,** p. 198. **113,** 10/15/36, p. 11. **114,** no. 5, 1909. **129,** pp. 180-182. **141,** pp. 20, 22, 34. **153,** pp. 53-54. **158. 159,** p. 130.

Rebate (203)

(See also Knob Hill.)

Loc: NE¼ sec. 22, (37-32E), Republic dist. **Owner:** Knob Hill Mines, Inc. (1943 ——). Rebate Mining Co. (1918). **Ore:** Gold, silver. **Ref: 68,** p. 11. **112,** p. 199.

Republic (Blaine Republic) (228)

Loc: NE¼ sec. 12, (36-32E). **Elev:** 3,300 to 3,900 ft. **Access:** ½ mi. SW. of Republic by road. **Prop:** 13 patented claims. **Owner:** Day Mines, Inc., Wallace, Idaho (1951 ——). Republic Gold Mining & Milling Co. (1898). Republic Consolidated Gold Mining Co. (1902). New Republic Co. (1909). Rathfon Reduction Works (1914-1926). Blaine-Republic Co. (1932-1936). Eureka Mining & Milling Co. (1937-1940). Aurum Mining Co. (1950). **Ore:** Gold, silver, selenium. **Ore min:** Pyrite, free gold. **Gangue:** Quartz, calcite. **Deposit:** Vein as much as 8 or 10 ft. wide, but av. 2 or 3 ft. Vein is composed principally of chalcedonic quartz. Concentrically crustified. Crustifications marked by dark crenulated bands. **Dev:** About 2 mi. of adits, drifts, winzes, raises, and crosscuts. **Assays:** Some handpicked ore assays $100, but most av. 0.14 oz. Au, 0.56 oz. Ag. **Prod:** Est. at $1,400,000 to 1910; 2,757 tons of ore to smelter 1937; shipped 1933-1946. **Ref: 1-A,** 1900, p. 419. **7,** pp. 154-157. **28,** pp. 87-91. **43,** vol. 68, pp. 725-726. **76. 88,** pp. 16-18. **97,** 1908-1918, 1921-1924, 1929, 1931, 1934, 1935, 1938-1941. **98,** 1918-1926. **99,** 11/27/34. **104,** 4/30/34; 6/30/36, p. 22; 12/15/36, pp. 26-27. **106,** 7/21/32, p. 7; 11/3/32, p. 9; 9/7/33; 10/5/33. **112,** pp. 164, 166, 198. **113,** 10/15/36, p. 11. **141,** p. 22. **153,** pp. 44-50. **158. 159,** p. 130.

Rosario (231)

Loc: Sec. 31, (32-37E) and sec. 36, (32-36E), Covada dist. **Access:** 28 mi. by road to railroad at Kettle Falls. **Prop:** 2 unpatented claims. **Owner:** R. V. Messenger, Inchelium, Wash. (1941). **Ore:** Gold, silver. **Deposit:** Crushed and mineralized country rock. **Dev:** 300-ft. adit, 50-ft. adit. **Ref: 158. 163,** p. 81.

San Poil (217)

(See also Day Mines.)

Loc: SW¼NE¼ sec. 34, (37-32E), Republic dist., on W. side of Eureka Gulch. **Access:** 1 mi. by road from Republic. **Prop:** 1 claim, S. of the Ben Hur. **Owner:** Day Mines, Inc., Wallace, Idaho (1951 ———). San Poil Consolidated Mining Co. (1910-1918). Western Union Mines Co. (1914-1915). West Hill Mining Co. (1915-1922). Western Silica & Sand Co. (1918). Trail Mining & Smelting Co. (1919). Zwang, Ivan and Co. (1933). Aurum Mining Co. (1950). **Ore:** Gold, silver. **Deposit:** Latite porphyry cut by a quartz-calcite vein thought to be the southward extension of the Ben Hur vein. Vein is cut by several faults. Ore in lenses up to 8 ft. thick. **Dev:** 2,500 ft. of adits which expose the vein for 1,000 ft. **Assays:** Some of the ore shipped av. 0.7 oz. Au, 4 oz. Ag. **Prod:** 1909-1921, 1931, 1935. 200 tons prior to 1902. **Ref: 7,** pp. 162-163. **33,** 1908, p. 1215. **97,** 1910-1921, 1931, 1935. **98,** 1918-1925. **104,** 12/15/34, p. 24. **106,** 11/5/31; 4/6/33. **112,** p. 201. **129,** p. 188. **153,** pp. 62-63. **158. 159,** p. 130.

Seattle (218)

Loc: W½NW¼ sec. 34, (37-32E), Republic dist. **Prop:** 1 patented claim: Seattle. **Owner:** J. A. Wiseman, Republic, Wash. (1939). **Ore:** Gold, silver. **Dev:** 60-ft. shaft. **Prod:** 1938, 1939. **Ref: 58,** p. 61. **97,** 1939, p. 490; 1940, p. 477. **158.**

South Penn (204)

(See also Day Mines.)

Loc: Sec. 22, (37-32E), Republic dist. **Owner:** Day Mines, Inc., Wallace, Idaho (1951 ———). C. Trevitt and E. Pierce, Republic, Wash., leasing from Aurum Mining Co. (1949). **Ore:** Gold, silver. **Gangue:** Quartz. **Assays:** 1945 production of 172 tons yielded $2,168. **Prod:** 1941-1949. **Ref: 22,** p. 11. **68,** p. 16.

Southern Republic

(See Princess Maude.)

Surprise (219)

(See also Day Mines.)

Loc: E½NE¼SE¼ sec. 34, (37-32E), on E. side of Eureka Gulch. **Access:** 1 mi. NW. of Republic by road. Lies between Quilp on S. and Pearl on N. **Prop:** 1 claim. **Owner:** Day Mines, Inc., Wallace, Idaho (1951 ———). Syndicated Deep Mines (1907-1908). Republic Mines Corp. (1910-1912). Western Union Mines Co. (1914-1915). West Virginia Mining Co. (1915-1918). Republic

Consolidated Mines Corp. (1915-1922). Northport Smelting & Refining Co. (1916-1926). Surprise Mining Co. (1924-1926). Aurum Mining Co. (1950). **Ore:** Gold, silver. **Ore min:** Gold, pyrite. **Gangue:** Quartz. **Deposit:** 4- to 8-ft. vein in propylitic quartz latite. Vein consists of banded quartz and included fragments of country rock. **Dev:** 1,100 ft. of drifts and crosscuts (1909). 700-ft. inclined shaft. **Assays:** 2,400 tons av. $21.65 per ton. **Prod:** Over $1,000,000. 1910-1923, 1934, 1938, 1947-1950. **Ref:** 7, pp. 159-160. 33, 1908, p. 1294. 43, 1900, p. 617. 68, p. 5. 88, pp. 19-20. 97, 1910-1923, 1934. 98, 1918-1926. 112, p. 211. 114, no. 5, 1909. 116, no. 11, 1907, p. 27; no. 12, 1907, p. 15; no. 5, 1908, p. 117; no. 6, 1908, p. 139. 129, pp. 185, 186. 141, pp. 22, 34. 153, pp. 54-56. 158.

Surprise (198)

(See also Morning Star.)

Loc: Near N. line NW¼ sec. 16, (40-34E). **Access:** 1½ mi. S. of Danville by road on S. side of Lone Ranch Cr. **Owner:** Morning Star Mining Co., Spokane, Wash. (1936-1952). Faithful Surprise Mining Co. (1915). **Ore:** Gold. **Ore. min:** Free gold. **Deposit:** 2-ft. quartz vein in serpentine with some black slate and diorite. Coarse yellow gold in the quartz. **Dev:** Adit more than 300 ft. long. **Assays:** Ore shipped av. $17 per ton. **Prod:** Has produced. **Ref:** 7, p. 200.

Tom Thumb (202)

(See also Day Mines.)

Loc: S½SE¼ sec. 15, (37-32E), at head of Granite Cr. **Elev:** 3,200 ft. **Access:** 4 mi. N. of Republic by road. **Prop:** 6 patented claims. **Owner:** Day Mines, Inc., Wallace, Idaho (1951 ——). New Republic Mining Co. (1908-1910). Midget Gold Mining & Milling Co. (1909-1918). Aurum Mining Co. (1939-1950). **Ore:** Gold, silver. **Deposit:** Quartz spread widely through shale. Lodes not well defined. One 8-ft. vein in andesite. **Dev:** Said to be about 1,600 ft. of underground development, including a 375-ft. shaft. **Assays:** Ore said to range from $10 to $15 per ton, though some shipments yielded $23 per ton. 8 or 9 oz. Ag. per oz. Au. **Prod:** 1908-1910, 1915, 1916, 1934, 1938. **Ref:** 7, p. 166. 68, p. 5. 88, p. 19. 97, 1908-1910, 1915, 1916, 1935. 106, 6/4/31. 112, p. 191. 114, no. 5, 1909. 153, pp. 64-65. 158. 159, p. 130.

Trade Dollar (220)

(See also Day Mines.)

Loc: E½SW¼ sec. 27, (37-32E). **Access:** About 2 mi. by road NW. of Republic. **Prop:** 1 claim adjoining Silver Dollar on the S. **Owner:** Day Mines, Inc., Wallace, Idaho (1951 ——). San Poil Consolidated Mining Co. (1910-1918). Curlew Mining Co. (1915).

Mountain Lion Consolidated Co. (1934). Aurum Mining Co. (1950). **Ore:** Gold, silver. **Deposit:** Quartz vein from 20 in. to 13 ft. wide. **Dev:** 300-ft. shaft with short drifts on 2 levels. **Assays:** 1,000 tons of ore av. $17 per ton with 10 to 12 oz. Ag. per ton. **Prod:** About $25,000 by 1934. **Ref:** 7, p. 163. 68, p. 5. 97, 1915, p. 569. 104, 1/30/33, p. 19. 158.

Valley (Golden Valley, Lame Foot) (200)

Loc: Sec. 6, (37-33E). **Access:** 5 mi. from Republic by road. **Prop:** 3 patented claims: Valley, Valley No. 1, Valley No. 2. **Owner:** I. G. and Everett Hougland, Republic, Wash. (1939 ——). **Ore:** Gold, silver, selenium. **Ore min:** Aurous selenide, pyrite, tetrahedrite. **Gangue:** Quartz, calcite. **Deposit:** Vein est. to av. 7 ft. wide and 1,200 ft. in length in andesite. **Dev:** 400-ft. incline and other workings on 3 levels total 2,500 ft. **Assays:** Typical assay shows 0.28 oz. Au, 1.0 oz. Ag. Analysis shows 0.0027% to 0.0061% Se. **Prod:** 1,994 tons prior to 1941, 5,800 tons 1942, 1,302 tons 1943, 1950. **Ref:** 28, pp. 77-81. 76. 97, 1940, p. 477; 1941, p. 473. 108, 7/39, p. 22; 2/42, pp. 13-16. 158.

Virginia

(See Morning Star.)

JEFFERSON COUNTY

Rustler Creek (2)

Loc: On Rustler Cr., in T. 25 N., R. 7 W. Possibly in sec. 31. **Access:** 1½ mi. by trail, 21½ mi. by road to U. S. 101 at Quinault Lk. **Prop:** 5 claims. **Owner:** M. H. and P. A. Mulkey, C. Slough, and M. and V. Oberg. **Deposit:** Quartz vein in slate. **Dev:** Open cut. **Assays:** $7.00 to $22.40 Au. **Ref: 158.**

KING COUNTY

Apex (Bondholders Syndicate) (54)

Loc: SW¼ sec. 34, (26-10E), near headwaters of Money Cr., Miller R. dist. **Elev:** 3,150 ft. **Prop:** 16 unpatented claims: Apex, Big Dyke, Campbell, Floe, Floe No. 2, Jumbo, Jumbo No. 2, May, Milwaukee, LaPorte No. 2, and others. **Owner:** Apex Gold Mines, Inc. (1905-1922, 1934-1943). Bondholders Syndicate Mining Co. (1923-1926). National Gold Corp. (1928). **Ore:** Gold, silver, copper, lead. **Ore min:** Arsenopyrite, pyrite, chalcopyrite, galena, arsenolite. **Deposit:** Quartz vein 2 to 6 ft. wide fills a continuous fissure in granodiorite. High-grade ore occurs in narrow streaks in the vein. **Dev:** Adits on 5 levels total 2,500 lineal ft. **Assays:** High-grade ore $20 to $80 Au. Low-grade $5 to $7 Au. 237 tons av. $41.07 per ton, the values chiefly in Au. Ore and conc. shipped in 1920 av. 21% to 26% As, 18 to 20 oz. Ag, 1½ to 2½ oz. Au, 4½%

to 6% Pb. **Prod:** 300 tons valued at $80,000 prior to 1901. Produced 1905, 1908, 1910, 1912, 1913, 1916-1920, 1926, 1928, 1936-1943. $300,000 total. **Ref:** 24. **33**, 1907. **63**, p. 39. **88**, p. 85. **97**, 1905, 1908, 1910, 1912, 1913, 1916-1920, 1923, 1926, 1928, 1938-1941. **98**, 1922, 1925, 1926. **99**, 1/29/35. **104**, 7/15/34, p. 26; 12/30/36, p. 28; 4/30-/32, pp. 3-4; 9/30/33, p. 17; 12/15/34, p. 22. **114**, no. 6, 1906, p. 79; 1907, pp. 61-62; no. 5, 1909. **116**, no. 4, 1908, p. 91. **117**, no. 8, 1922, p. 5. **129**, pp. 301-305. **130**, p. 59. **141**, p. 22. **158**. **159**, p. 135.

Beaverdale (56)

Loc: Secs. 8 and 9, (25-10E), ½ mi. up Illinois Cr. from N. Fk. of Snoqualmie R. Vein crops out in canyon which enters Illinois Cr. from the W. **Elev:** 3,000 to 3,720 ft. **Access:** Trail. **Prop:** 6 patented claims. **Owner:** A. S. Ryland, Seattle, Wash. (1947). **Ore:** Gold. **Ore min:** Pyrite, arsenopyrite. **Gangue:** Quartz, gouge. **Deposit:** Altered and brecciated granodiorite along narrow zone of faulting in which are seams of quartz and sulfides up to 20 in. thick. **Dev:** 3 adits; one at 3,370 ft. is caved at portal, another at 3,610 ft. is 55 ft. long, and the other at 3,720 ft. follows vein for 140 ft. **Assays:** Said to run as high as $161 Au. **Ref: 11-A**, pp. 210-213. **158**.

Bondholders Syndicate

(See Apex.)

Carmack (60)

Loc: Secs. 7 and 8, (22-11E), on S. Fk. of Snoqualmie R. near Snoqualmie Pass. **Prop:** 5 claims. **Owner:** Carmack Gold & Copper Mining Co. (1902-1918). **Ore:** Gold, silver, lead, copper. **Deposit:** 3 veins 12, 2½, and 1 ft. wide. **Dev:** 375 ft. of tunnel and shafts. **Assays:** 1 oz. to 1½ oz. Au. **Prod:** 20 tons shipped prior to 1901. **Ref:** 33, 1907, p. 447. 88, p. 86. 98, p. 66. **158.**

Coney Basin (55)

Loc: Sec. 13, (25-10E) and sec. 19, (25-11E). **Elev:** 2,100 to 5,800 ft. **Access:** 8 mi. up Miller R. from its mouth by road and 2 mi. by trail. **Prop:** 15 claims: Coney Nos. 1 to 3, Jeanette Nos. 1 to 3, Black Bear, Black Bear Nos. 1 and 2, Jake Nos. 1 and 2, Leona Nos. 1 and 2, Gold Chest, Gold Chest No. 1. **Owner:** Coney Basin Gold Mines, Inc. leasing to John Whitham, Seattle, Wash. (1950-1952). **Ore:** Gold, silver, copper, zinc, lead. **Ore min:** Chalcopyrite, galena, sphalerite, pyrite, tetrahedrite. **Gangue:** Quartz. **Deposit:** Small persistent quartz veinlets along joint planes in granodiorite. Also 1 silicified mineralized zone 4 ft. wide. **Dev:** 1,650-ft. adit, another shorter adit, and some open cuts total about 3,000 ft. **Assays:** Av. of 22 samples gave 0.38 oz. Au, 11.97 oz. Ag. 3 other assays showed 0.24% to 1.30% Cu, 2.6% to 6.9% Pb, 2.9% to 9.1% Zn. 8 tons shipped to smelter in 1941 had 0.86 oz. Au,

19.71 oz. Ag, 0.82% Cu, 6.0% Pb, 6.0% Zn, 1.52% As, 0.26% Sb. **Prod:** 40 tons in 1895, produced in 1934, 1937-1939, 1941. **Ref: 46,** pp. 159-160. **63,** p. 36. **97,** 1935, 1936, 1939, 1940. **104,** 9/31/34, p. 22; 6/15/35, p. 28. **114,** no. 5, 1909. **132,** p. 87. **147,** pp. 154-185. **158. 159,** p. 135.

Damon and Pythias (53)

Loc: Sec. 33, (26-10E), at head of Money Cr., adjoining Apex property. **Elev:** 3,000 ft. **Access:** 1 mi. from end of Apex road, by which it is 6 mi. to railroad at Miller R. **Prop:** 18 unpatented claims. **Owner:** W. J. Priestley, Skykomish, Wash. (1939-1944). National Gold Corp. (1928-1932). **Ore:** Gold, silver, lead. **Ore min:** Arsenopyrite, pyrite, chalcopyrite, galena. **Gangue:** Quartz. **Deposit:** 2 veins, one of which is said to av. 3 ft. in width over distance of more than 900 ft. in granodiorite. **Dev:** 3,000 ft. of crosscut and drifts expose the veins for about 1,500 ft. of length and from 800 to 1,400 ft. of depth. **Improv:** Cookhouse and bunkhouse, 2,300 ft. of track. **Assays:** 23 tons shipped said to run 0.87 oz. Au, 9 oz. Ag, 4% Pb. Mine-run ore said to carry 7.86% As, 0.245 oz. Au, 2.2 oz. Ag. **Prod:** 23-ton shipment reported prior to 1940. **Ref: 63,** p. 39. **97,** 1928, p. 700. **104,** 4/30/32, pp. 3-4. **106,** 9/17/28, pp. 29-30; 5/1/30, p. 20. **157. 158.**

Lennox (59)

Loc: Secs. 7 and 18, (25-10E) and sec. 13, (25-9E), Buena Vista dist. **Elev:** 1,830 to 2,870 ft. **Access:** 23 mi. by road from railroad at North Bend. **Prop:** 45 claims held by possessory title. **Owner:** Priestley Mining & Milling Co. (1940 ——) leasing from Lennox Mining & Development Co., Seattle, Wash. **Ore:** Gold, silver, zinc, lead, copper, arsenic, antimony. **Ore min:** Sphalerite, chalcopyrite, pyrite, arsenopyrite, galena. **Gangue:** Quartz, calcite. **Deposit:** Shear zones in granodiorite are persistent to depths and lengths of several hundred ft., but mineralization is irregular. **Dev:** 5 adits and 2 open cuts. Lower crosscut 670 ft. long. 3 diamond drill holes totaling 1,240 ft., 120-ft. drift, 20-ft. drift. **Improv:** 4 bunkhouses, messhall, sawmill, 2 machine shops, a Pelton water wheel, and an 80 KW. generator (1950). **Assays:** A 1-ton lot of picked ore (1938): 1.14 oz. Au, 10.42 oz. Ag, 1.5% Cu, 1.2% Pb, 8.3% Zn, 6.18% As, 0.67% Sb. 8 samples assayed in 1947 ranged from $7.51 to $63.91 Au, Ag, Pb, Zn, Cu. **Prod:** 1-ton test shipment in 1947. **Ref: 11-A,** pp. 213-222. **133,** p. 38. **157. 158.**

Lucky Strike (57)

Loc: Secs. 9 and 10, (25-10E), Buena Vista dist. **Elev:** 3,000 to 4,000 ft. **Access:** 2-mi. trail from N. Fk. road about 3½ mi. above mouth of Lennox R. **Prop:** 3 unpatented claims. **Owner:** George and Gary Rutherford, Seattle, Wash. (1947-1951). **Ore:** Gold, silver, copper. **Ore min:** Pyrite, arsenopyrite, chalcopyrite. **Gan-**

gue: Quartz. **Deposit:** Said to be an 8-ft. ledge traceable for 500 ft. Two parallel faults about 6 ft. apart in granodiorite containing quartz-sulfide seams as much as 12 in. thick. **Dev:** 30-ft. shaft. **Assays:** Reportedly assays $20 to $30 Au, Ag, Cu. **Ref: 11-A,** pp. 207-210. **158.**

Monte Carlo (58)

Loc: NW¼ sec. 4, (25-10E), near head of N. Fk. of Snoqualmie R. **Elev:** 3,440 ft. **Access:** 4 mi. by trail from road at Lennox camp. **Prop:** 13 unpatented claims. **Owner:** E. P. Courtney and Morton Van Nuys, Seattle, Wash. (1947). **Ore:** Gold, silver. **Ore min:** Pyrite, malachite, arsenopyrite, molybdenite. **Gangue:** Quartz, tourmaline. **Deposit:** Quartz vein along a shear zone in granite varies from 1 in. to 20 in. wide. **Dev:** 340-ft. drift and a crosscut. 2 other adits reported. **Improv:** Cabin (1947). **Assays:** Said to assay $25 to $55 Au, Ag. **Ref: 58,** p. 45. **158.**

Pythias

(See Damon and Pythias.)

San Jose (61)

Loc: SE¼ sec. 27, (22-10E), Cedar R. dist. **Ore:** Gold, copper. **Assays:** 10 tons shipped in 1894 av. $12 per ton. **Ref: 63,** pp. 47-48.

White River (62)

Loc: Approx. sec. 6, (19-8E), on S. side of Quartz Mtn. **Ore:** Gold. **Deposit:** Silicified and altered volcanic rocks associated with the alunite deposits. **Dev:** 2 adits, one 300 ft. long. **Improv:** About 1890 a stamp mill was erected to mill the ore but was soon abandoned. **Assays:** Said to range from 54¢ to $1.40 Au. **Ref: 158.**

KITTITAS COUNTY

Aurora (Lynch, Paramount) (122)

Loc: Secs. 26 and 27, (24-14E), on E. side of Cle Elum R. **Access:** 24 mi. by road and trail from railroad at Roslyn. **Prop:** 16 patented claims. **Owner:** John Lynch, Yakima, Wash., leasing to Paramount Mines, Inc. (1940). **Ore:** Gold, silver, copper. **Ore min:** Free gold, arsenopyrite. **Deposit:** 4-ft. quartz vein. **Dev:** 2,000-ft. haulage adit and 2 shafts each more than 200 ft. deep. **Assays:** Ore said to assay 1 oz. Au, 14 oz. Ag, 6% Cu, 28% As, probably on a picked sample. **Prod:** Has produced. **Ref: 13,** p. 133. **63,** pp. 61-62. **111,** p. 7. **158.**

Cascade Chief (Morrison, First of August, Gladstone) (157)

Loc: SE¼SW¼ sec. 26, (21-17E), 4 mi. up Cougar Gulch from Liberty Post Office. **Elev:** 3,400 ft. **Access:** Road. **Prop:** 5 lode claims: First of July, First of August, Fourth of July, The Gladstone, Uncle Sam. Also one placer claim: Cougar. **Owner:** Cascade Chief, Inc. (1934-1936). State of Washington Mining Co. (1915). **Ore:** Gold. **Ore min:** Free gold. **Deposit:** 3 shear zones in sandstone av. 4 ft. in width and carry stringers of mineralized quartz. **Dev:** Crosscut and some old workings total several hundred ft. of workings. **Assays:** Channel sample across the vein assayed $7.24 Au. One assay showed $129.35 Au. **Prod:** 1911, 1938, 1939. **Ref: 97,** 1911, p. 786. **104,** 4/30/34, p. 22; 11/30/36, p. 30. **158.**

Clarence Jordin (162)

Loc: Sec. 2, (20-17E), on Snowshoe Ridge, Swauk dist. **Prop:** 3 unpatented claims, including Ace of Diamonds. **Owner:** W. L. Palmer, Electric City, Wash. (1941). **Ore:** Gold, silver. **Ore min:** Free gold. **Gangue:** Quartz, calcite. **Prod:** $35,000 reported. Produced from Ace of Diamonds claim in 1952. **Ref: 58,** p. 35. **106,** 4/21/32, p. 13. **108,** 1/53, p. 90. **158.**

Cougar (158)

Loc: Sec. 36, (21-17E), Swauk dist. **Elev:** 3,500 ft. **Owner:** Washington Gold Co., Inc., Tieton, Wash. (1938). State of Washington Mining Co. (1915). Cougar Gulch Mining Co. (1934). **Ore:** Gold. **Ore min:** Free gold. **Deposit:** Quartz vein in a decomposed basaltic dike. **Prod:** Test shipment in 1938. **Ref: 143,** p. 80. **144,** p. 9. **158.**

Ewell (Flag Mountain) (161)

Loc: N½ sec. 1, (20-17E), Liberty area. **Owner:** Ellery Lamb et al., Seattle, Wash. (1953 ——). William L. Ewell, Cle Elum, Wash. (1952). **Ore:** Gold. **Improv:** 10-ton mill (1953). **Ref: 108,** 1/53, p. 90. **133,** p. 32.

First of August

(See Cascade Chief.)

Flag Mountain

(See Ewell.)

Flodine (159)

Loc: Sec. 25, (21-17E) and sec. 30, (21-18E). **Prop:** 12 unpatented lode claims: Ohio, Cougar, Red Rust Nos. 1 to 3, Gold King, Comet, Venus, Meteor, Jupiter, Mars, North Star; and two placer claims: Bull Moose, Morning Star. **Owner:** True Fissure Gold Mines (1928). **Ore:** Gold. **Ore min:** Free gold, pyrite. **Gangue:** Quartz. **Deposit:** Fissured zone. **Dev:** 260-ft. shaft, 288-ft. adit, 140-ft. crosscut with 159-ft. drift. **Assays:** $6.00 to $17.20 Au. **Prod:** Several thousand dollars from oxidized zone prior to 1928. **Ref: 158.**

Gladstone

(See Cascade Chief.)

Golden Fleece (Mercer, T-Bone) (155)

Loc: NE¼SW¼ sec. 13, (21-17E), Swauk dist. **Prop:** 3 claims and a 5-acre mill site. **Owner:** Oren F. Fry, Mukilteo, Wash. (1941). Liberty Mines, Inc. (1934-1935). **Ore:** Gold, silver. **Ore min:** Pyrite, free gold. **Gangue:** Quartz, calcite. **Deposit:** Mineralized shear zone about 4 in. wide cuts carbonaceous shales. **Dev:** 100-ft. adit and open cut. **Assays:** Tr. to 0.34 oz. Au. **Prod:** $30,000 reported. **Ref: 58,** pp. 25, 43. **104,** 10/15/35, p. 24; 11/15/35, p. 24. **158.**

Ida Elmore (125)

Loc: S½ sec. 24 and NE¼ sec. 25, (23-14E), Cle Elum dist. **Prop:** 2 patented claims: Ida Elmore, Apex. **Owner:** Meleo S. Pechet, Seattle, Wash. (1949-1952). **Ore:** Gold, silver. **Deposit:** Quartz vein about 12 to 18 in. wide in serpentine. **Dev:** 235 ft. of crosscut and about 100 ft. of drift, also some open cuts. **Assays:** Av. ¾ to 1 oz. Au. **Ref: 13,** p. 134. **63,** p. 64. **146,** p. 14. **158.**

Liberty (160)

Loc: NW¼NE¼ sec. 25, (21-17E) and sec. 19, (21-18E), in Lyons Gulch. **Access:** Road. **Prop:** 12 claims: Contact No. 1, Orogrande, Orogrande No. 1, Big Boy Nos. 1 and 2, Big Chief Nos. 1 and 2, Golden Eagle, Sunset Nos. 2 and 3; and 2 fractions. **Owner:** A. C. Pellard, Mount Vernon, Wash. (1941). **Ore:** Gold. **Ore min:** Pyrite. **Deposit:** Shear zone about 4 ft. wide in carbonaceous shale contains discontinuous mineralized stringers of quartz and calcite. **Dev:** 1,200-ft. adit with a 145-ft. drift, and another 200-ft. adit. **Improv:** 2 log cabins (1952). **Assays:** Test run on 33 tons of ore returned 10 oz. Au, or 0.3 oz. per ton. **Prod:** 1935-1936. **Ref: 58,** p. 39. **97,** 1937, p. 551. **158.**

Lynch

(See Aurora.)

Maud O. (126)

Loc: NW¼ sec. 25, (23-14E), Cle Elum dist. **Ore:** Gold. **Ore min:** Pyrite, arsenopyrite. **Deposit:** Mineralized quartz vein in a zone of crushed greenstone into which serpentine has been intruded. Vein irregular. **Prod:** 1930. **Ref: 58,** p. 42. **63,** p. 64. **100,** 1897, pp. 97, 132, 158. **146,** p. 14.

Mercer

(See Golden Fleece.)

Morrison

(See Cascade Chief.)

Mountain Daisy (163)

Loc: Sec. 1, (20-17E) and sec. 6, (20-18E), Swauk dist. **Owner:** Ollie Jordan, lessor, Rt. 2, Cle Elum, Wash. (1952). **Ore:** Gold.

Prod: 1934-1938. **Ref: 97,** 1937, p. 551; 1938, p. 459; 1939, p. 490. **133,** p. 37.

Ollie Jordin (164)

Loc: Sec. 2, (20-17E), ¾ mi. up Williams Cr. from Liberty, Swauk dist. **Owner:** Ollie Jordin and Miss Ollie Blissett (1938). **Ore:** Gold. **Ore min:** Free gold. **Deposit:** Silicified zone 4 ft. wide in Swauk sandstone contains small quartz-calcite stringers in which "wire" gold occurs. **Dev:** 170-ft. adit. **Assays:** Gold occurs in rich pockets, hence assays are erratic. Ore av. $40 per ton on 500 tons mined. **Prod:** About $20,000 in gold during 2 years prior to 1934. **Ref: 158.**

Paramount

(See Aurora.)

Silver Bull (123)

Loc: Contiguous to Fish Lk. in Stevenson Gulch, Cle Elum dist. **Prop:** 1 claim. **Owner:** James Greeve, E. P. Cassman, and August Sassi (1892). **Ore:** Gold, silver. **Ore min:** Pyrite. **Deposit:** 5-ft. vein of white quartz carrying pyrite and free-milling ore. **Dev:** 4 adits 40, 70, 90, and 130 ft. long. **Assays:** $100 Au, Ag. **Ref: 13,** p. 134.

Silver Creek (124)

Loc: Sec. 12, (23-14E), Fish Lk. area. **Access:** 24 mi. by road to railroad at Ronald. **Owner:** Cle Elum River Mining Co., Inc., Seattle, Wash. (1952 ——). Silver Creek Mining Co., Tacoma, Wash. (1939). **Ore:** Gold, silver. **Deposit:** Quartz vein 15 to 20 ft. wide reported, but av. values are too low to mine whole width, although some assays show good values in gold and silver. **Dev:** 5 adits, 1 shaft. **Assays:** A 5-ton shipment gave $12 per ton. **Prod:** 1937, 1939. **Ref: 58,** p. 52. **97,** 1938, p. 459; 1940, p. 478. **158.**

T-Bone

(See Golden Fleece.)

Wall Street (156)

Loc: Sec. 30, (21-18E), 5½ mi. up Cougar Gulch from Liberty. **Access:** 1½ mi. trail. **Prop:** 3 unpatented claims. **Owner:** Wm. Newstrum, Ellensburg, and R. J. Jordan, Liberty, Wash. (1934). Wall Street Mining Co. (1909-1915). **Ore:** Gold. **Gangue:** Quartz. **Deposit:** Silicified fracture zones in sandstone. **Dev:** 900-ft. adit connected by a raise with a 306-ft. adit 260 ft. above the lower adit. A 54-ft. winze in the upper adit. **Assays:** $5.60 to $10.50 Au. **Prod:** $50,000 reported prior to 1935. 1938. **Ref: 58,** p. 71. **97,** 1939, p. 490. **104,** 4/30/34, p. 22. **158.**

OKANOGAN COUNTY

Alder (94)

Loc: Secs. 25, 26, 35, and 36, (33-21E). **Elev:** 3,000 to 3,800 ft. **Access:** 5 mi. SW. of Twisp by road. **Prop:** 3 patented and 17 unpatented claims. **Owner:** Alder Gold-Copper Co., Spokane, Wash. (1949 ——). **Ore:** Gold, zinc, copper, silver. **Ore min:** Chalcopyrite, pyrite, sphalerite, native copper, pyrrhotite. **Deposit:** Silicified zone 15 to 75 ft. wide in sheared argillite. **Dev:** 3 adits total several hundred ft. Large open pit. **Improv:** 300-ton flotation mill (1952). **Assays:** 6,831 tons shipped in 1939 av. 0.55 oz. Au, about 0.50 oz. Ag, 0.16% to 0.55% Cu. **Prod:** 1937; 1939, 13,000 tons of ore; 1940, 9,000 tons of ore; 1941; 1942, about 4,000 tons; 1950; 1951. **Ref: 33,** 1918, p. 1454. **97,** 1928, p. 700. **104,** 8/15/32, p. 29; 1/15/37, p. 28. **106,** 12/19/29; 10/2/30. **108,** 11/39, p. 2; 9/41, p. 36. **133,** p. 29. **158.**

Alice (186)

(See also Summit, Black Bear, War Eagle.)

Loc: Sec. 36, (39-25E) and sec. 31, (39-26E), Palmer Mtn. dist. **Prop:** 3 claims: Black Bear, Summit, War Eagle. **Owner:** R. D. Heffernan, Tonasket, Wash. (1949). **Ore:** Gold. **Ref: 52,** p. 5.

Allison

(See Okanogan Free Gold.)

American Flag (Oriental and Central) (87)

Loc: SE¼ sec. 30, (36-20E), on a cliff about 1 mi. NE. of Mazama. **Elev:** 2,600 to 3,000 ft. **Access:** ½ mi. by trail and 1½ mi. by road from Mazama. **Prop:** 2 patented claims, 1 unpatented claim. **Owner:** Mahlon McCain and associates (1946). **Ore:** Gold, copper, zinc, silver. **Ore min:** Pyrite, arsenopyrite, chalcopyrite, sphalerite. **Gangue:** Quartz, calcite. **Deposit:** 2 mineralized fault zones 2 to 40 in. wide in diorite contain small quartz stringers and sulfides. **Dev:** 2 adits and a sublevel together with a raise and winze total about 1,400 ft. **Assays:** 0.16 to 3.08 oz. Au across widths of 2 in. to 2 ft. **Prod:** A few hundred tons produced before 1910, and small production in 1940. **Ref:** 40, p. 14. **88,** p. 35. **97,** 1941, p. 474. **106,** 3/19/31, 4/2/31. **108,** 10/39, p. 32; 7/40, p. 38. **157. 158.**

American Rand

(See Spokane.)

Bellevue (177)

Loc: NW¼ sec. 4, (39-26E), Wannacut Lk. dist. **Prop:** 5 claims. **Ore:** Gold, silver, copper. **Ore min:** Arsenopyrite, pyrite, chalcopyrite, pyrargyrite, stephanite, a little native silver, free gold, and possibly gold-silver telluride. **Deposit:** Quartz vein from 10 in. to 3 ft. wide enclosed in slate. Vein av. 15 in. wide. **Dev:** Small shaft and several open cuts. **Assays:** A test shipment of 1,000 lb. returned $75 Au, Ag, more than half of which was Au. **Prod:** Several tons. **Ref: 63,** p. 100. **154,** p. 101.

Black Bear (184)

(See also Alice.)

Loc: NE¼ sec. 36, (39-25E), on Palmer Mtn. **Elev:** 2,500 ft. **Prop:** Part of Alice property. **Owner:** R. D. Heffernan, Tonasket, Wash. (1949). **Ore:** Gold, silver, copper. **Ore min:** Free gold, pyrite. **Deposit:** A 4-ft. quartz vein along contact of chlorite schist and serpentine. Vein carries a 3-ft. paystreak. **Dev:** 2,500 ft. of underground workings. **Assays:** Av. $18 Au, Ag, Cu. **Prod:** $150,000 prior to 1902; 77 tons in 1947. **Ref: 12,** p. 62. **13,** p. 99. **63,** p. 98. **88,** p. 31.

Bodie (Northern Gold) (195)

Loc: SW¼ sec. 3, (38-31E) and sec. 34, (39-31E), on Toroda Cr., N. of Wauconda. **Elev:** 3,000 ft. **Access:** 22 mi. by road to railroad at Republic. **Prop:** 5 patented, 10 unpatented claims. **Owner:** Toroda Gold Mines Corp. subleasing from Northern Gold Corp. (1939-1941). Bodie Transportation & Mining Co. (1907). B and F Mining Co. (1908-1909). Duluth-Toroda Mining Co. (1910-1912). Toroda Development Co. (1915-1916). Northern Gold Corp. (1934-1941). **Ore:** Gold, silver. **Ore min:** Free gold, pyrite. **Deposit:** Finely disseminated ore in quartz-calcite vein. Vein much altered and leached near surface, and is similar to those in Republic dist. Veins are in tuff and andesite. One vein av. 4 ft. wide. **Dev:** Several thousand ft. of workings on 4 levels. **Assays:** 302 tons produced 1940-1944 returned $9,770 or $32.35 per ton. Ore said to av. about $10.00 Au. **Prod:** 1906, 1907, 1909-1911, 1914, 1915, 1934-1944. 66,032 tons 1935-1944. **Ref: 58,** pp. 9, 68. **97,** 1907-1912, 1915, 1916, 1934-1944. **99,** 11/27/34, 1/8/35, 1/29/35, 2/12/35. **104,** 1/30/35, p. 23; 1/30/36, p. 22; 12/30/36, p. 29. **106,** 10/34. **116,** no. 3, 1909, p. 230. **117,** no. 24, 1919, p. 12. **158.**

Bullfrog (174)

Loc: SW. part sec. 33, (40-26E), Palmer Mtn. dist. **Prop:** 18 claims. **Owner:** Bullfrog Gold Mining Co. (1902). **Ore:** Gold, silver. **Ore min:** Pyrite, black metallic sulfide mineral. **Deposit:** Reportedly a 7-ft. quartz vein in quartzite and sericitic schist traceable for 3,000 ft. **Dev:** Adit, 160-ft. shaft, 140-ft. shaft. **Assays:** 10-ton test yielded $17 per ton, of which $12 was in gold and $5 in silver (1902). **Prod:** 4,600 lb. shipped. **Ref: 63,** p. 100. **88,** pp. 30-31. **154,** p. 101.

Butcher Boy (188)

Loc: SW¼NW¼ sec. 21, (40-30E), Myers Cr. dist. **Access:** Road from Chesaw. **Prop:** 2 claims. **Owner:** Leased by James

Miller, Chesaw, Wash. (1938). Butcher Boy Gold Mining Co. (1908-1915). **Ore:** Gold, silver, lead, zinc. **Ore min:** Pyrite, pyrrhotite, sphalerite, galena. **Gangue:** Quartz, calcite. **Deposit:** Quartz vein in argillite from 1 in. to 6 ft. wide. Argillite has been intruded by granite. **Dev:** 326-ft. adit, shaft, stope. **Assays:** Vein in adit said to assay $7 to $17 Au, 70¢ Ag. Some ore said to carry as much as 7 oz. Au. **Prod:** 11 carloads of ore 1907. **Ref:** 46, p. 175. 154, pp. 49-50.

Campbell

(See Holden-Campbell.)

Caribou (191)

Loc: SE¼ sec. 14, (40-30E), Myers Cr. dist. **Owner:** British Columbia Copper Co. (1908-1916). **Ore:** Gold, silver, copper. **Prod:** 8 carloads of ore in 1916. **Ref:** 33, 1908, p. 413. 97, 1916, p. 613; 1917, p. 503.

Central

(See American Flag.)

Chelan (Pennington) (111)

Loc: Near NE. cor. sec. 17, (30-22E), near head of Squaw Cr. on N. side of the valley. **Elev:** 3,000 ft. **Access:** Road up Squaw Cr. **Prop:** At least 7 claims. **Owner:** Roy Pennington (1943). **Ore:** Gold, tungsten. **Ore min:** Scheelite, pyrite, free gold. **Deposit:** Several mineralized quartz veins from 6 to 24 in. wide along shear zones in diorite. **Dev:** Caved adit on No. 7 vein and 40-ft. adit on No. 4 vein, also some open cuts. **Assays:** Gold values of $21 and $161 are reported. Scheelite content less than ¼%. **Ref:** 37, p. 49.

Chloride Queen (168)

Loc: SE¼NE¼ sec. 36, (40-25E), on SW. slope of Ellemeham Mtn., Nighthawk dist. **Owner:** Mr. Everett, Nighthawk, Wash. (1939). **Ore:** Gold. **Ore min:** Pyrite, free gold. **Deposit:** A 1- to 4-ft. iron-stained quartz vein in argillite is sparsely mineralized with pyrite and reportedly free gold. **Dev:** 50-ft. inclined shaft and a drift. **Prod:** 1936, 1937. **Ref:** 97, 1937, p. 552; 1938, p. 459. 158.

Continental

(See Mazama Queen.)

Crescent

(See Triune.)

Crown Point

(See Imperial.)

Crystal Butte (Mother Lode) (193)

Loc: Center W½ sec. 35, (40-30E), Myers Cr. dist. **Access:** Road from Chesaw. **Prop:** 6 claims. **Owner:** Being purchased on lease and option by R. C. Hirst from Arthur and Joseph Little, Chesaw, Wash. (1938). **Ore:** Gold, silver, lead, zinc, copper. **Ore min:** Galena, chalcopyrite, sphalerite, pyrite, arsenopyrite. **Deposit:** A mineralized quartz vein av. 1 ft. in width occurs along the contact of limestone and argillite. **Dev:** Inclined shaft, adit, open cuts, some old caved workings. **Assays:** A carload and a truckload shipped in 1937 netted $40 per ton in gold and silver. **Prod:** A carload and truckload in 1937. Produced 1938-1941. **Ref:** **97,** 1938-1941. **154,** pp. 48-49. **158.**

Denver City

(See Leadville.)

Friday (Tom Hal) (107)

Loc: SW¼ sec. 20, (30-23E), Squaw Cr. dist. **Access:** Paved highway from Pateros and cable car across Methow R. **Prop:** 5 claims. **Owner:** J. J. Sullivan, Pateros, Wash. (1942). **Ore:** Gold, silver. **Ore min:** Pyrite, chalcopyrite, arsenopyrite, bornite, malachite. **Deposit:** 1-ft. quartz vein in gneissic diorite. **Dev:** 110-ft. crosscut, 100-ft. winze, 240 ft. of drift (1902). **Assays:** 10 tons selected ore yielded $70 per ton prior to 1902. Other assays showed 0.56 oz. Au, 1.3 oz. Ag. **Prod:** $5,000 prior to 1897, 1 carload in 1940. **Ref:** **58,** p. 68. **63,** p. 86. **88,** p. 39. **158.**

Gold Axe (192)

Loc: SW¼ sec. 24, (40-30E), Myers Cr. dist. **Owner:** W. C. Carpenter, Grand Coulee, Wash. (1938). Buckhorn Mining Co. (1935). **Ore:** Gold, silver, copper. **Prod:** 16 or 17 carloads (1914-1915). **Ref:** **58,** p. 24. **97,** 1911, 1915, 1916, 1928, 1934, 1935. **104,** 10/15/35, p. 25; 11/15/35, p. 24; 3/30/38. **106,** 12/1/32.

Gold Coin (98)

Loc: E½ sec. 5, (30-22E), Squaw Cr. dist. **Prop:** 2 patented claims: The Billy, Gold Coin No. 1. **Owner:** Joe Douglas, Methow, Wash. (1949). Gold Coin Mining Co. (1932). **Ore:** Gold. **Prod:** 1934, 1941. **Ref:** **97,** 1935, p. 354. **104,** 2/29/32, p. 30. **158.**

Gold Crown (171)

Loc: Sec. 31, (39-26E), Palmer Mtn. dist. **Ore:** Gold. **Gangue:** Quartz. **Deposit:** 10-ft. vein. **Assays:** Dump sample showed $105 Au. **Prod:** Amount not known. **Ref:** **12,** p. 61. **13,** p. 100.

Gold Crown (167)

Loc: SW¼ sec. 32, (40-25E), Chopaka area. **Access:** 8 mi. by road to Loomis. **Prop:** 3 claims. **Owner:** Victor Byrski, Vern C. Anderson, Tacoma, Wash., and Perry O. Luce, Coulee Dam, Wash.

(1951). **Ore:** Gold, silver, copper. **Ore min:** Pyrite, chalcopyrite. **Gangue:** Quartz. **Deposit:** Veins a few inches to 4 ft. thick in granite. **Dev:** 60-ft. shaft and shallow pits. **Assays:** Av. 2 oz. Au, 4 oz. Ag, 4% Cu reported. **Ref: 150,** p. 32. **158.**

Gold Crown

(See Spokane.)

Gold Key (88)

Loc: Near NW. cor. sec. 30, (36-20E), about 1,000 ft. NW. of upper workings of Mazama Pride, Mazama dist. **Elev:** 2,500 ft. **Prop:** Part of Mazama Pride group (?). **Owner:** Alva Sharp, Mazama, Wash. (1946). **Ore:** Gold, copper. **Ore min:** Pyrite, arsenopyrite, and some chalcopyrite. **Deposit:** Quartz stringer in diorite mineralized to some extent. **Dev:** 110-ft. adit with a 15-ft. winze. **Assays:** 37 tons shipped av. 0.7 oz. Au. **Prod:** 37 tons of ore shipped in 1931. **Ref: 158.**

Golden Zone (166)

Loc: Near SE. cor. sec. 7, (40-25E), at base of Mt. Chopaka, Nighthawk dist., 2 mi. S. of the international boundary. **Elev:** 1,500 ft. **Prop:** 5 patented claims. **Ore:** Gold, silver, lead, copper, reportedly molybdenum. **Ore min:** Pyrite, chalcopyrite, free gold, argentiferous galena, occasionally sphalerite and arsenopyrite. **Gangue:** Quartz. **Deposit:** Vein up to 4 ft. thick in granite a short distance from its contact with metamorphic rocks. **Dev:** 5,000 ft. of adits and drifts. **Assays:** Up to $10 Pb, $25 Au, 4 oz. Ag. **Prod:** Prior to 1911, 1939. **Ref: 63,** p. 103. **88,** p. 30. **97,** 1915, p. 572; 1940, p. 478. **145,** p. 95. **154,** p. 95.

Grand Summit

(See Palmer Summit.)

Grandview

(See Leadville.)

Gray Eagle (189)

Loc: Sec. 16, (40-30E), Myers Cr. dist. **Access:** Road from Chesaw. **Owner:** C. N. Bagwell, M. L. Pierce, Chas. Atchison, lessees, Loomis, Wash. (1941). Patrick Welch estate, Spokane, Wash. (1937). **Ore:** Gold. **Ore. min:** Pyrite, free gold. **Deposit:** Small quartz veinlets filling fractures in altered granitic rock. **Prod:** $8,000 in 1916. 1936-1939. **Ref: 58,** p. 27. **97,** 1916, 1921, 1923, 1924, 1935, 1937-1940. **104,** 10/30/34, p. 22. **113,** 4/15/37, p. 7. **158.**

Grubstake (100)

(See also Holden-Campbell.)

Loc: NE¼SW¼ sec. 10, (30-22E), Squaw Cr. dist. **Access:** 1½ mi. up Vinegar Lk. trail from Squaw Cr. road. **Prop:** Part of Holden-Campbell group. **Owner:** S. J. Holden and A. C. Campbell, Chelan, Wash. (1943). **Ore:** Gold. **Ore min:** Pyrite. **Deposit:** Quartz vein in gneissic granitic rock. **Dev:** Adit, caved (1938). **Ref:** 158.

Hiawatha (Josie) (178)

Loc: Near center N½ sec. 10, (39-26E), ¼ mi. NW. of and 300 ft. above the Triune mine, Wannacut Lk. dist. **Elev:** 3,200 ft. **Access:** Good road from Oroville. **Prop:** 6 claims, 3 of which are patented. **Owner:** Dell Hart & Sons, Oroville, Wash. (1938). Hiawatha Mining Co. (1934). **Ore:** Gold, silver, lead, zinc, copper. **Ore min:** Auriferous galena, chalcopyrite, pyrite, sphalerite. **Deposit:** 4 quartz veins in argillite. The Hiawatha vein has been most developed. It av. 3 ft. in width and is traceable for 2,500 ft. Mineralization sparse. **Dev:** Two 80-ft. adits 80 ft. apart and connected by drift near the face. **Assays:** Av. of 22 assays was $26 Au, 1 oz. Ag. One sample showed 9.5% Pb, 0.32 oz. Au, 61.7 oz. Ag. **Prod:** 1938. **Ref:** 46, p. 226. **54**, p. 3. **97**, 1934, p. 297; 1935, p. 353; 1939, p. 490. **154**, p. 99. **158**.

Hidden Treasure (Sunshine, Triangle) (104)

(See also Highland.)

Loc: SE¼ sec. 11, (30-22E), Squaw Cr. dist. **Access:** 2 mi. of steep truck road from Squaw Cr. road. **Prop:** 6 claims. **Owner:** Part of the property owned by the Highland Mining & Milling Co. (1949). Hidden Treasure Mining & Milling Co. (1902-1909). Hidden Treasure Gold Mining Co. (1910). Triangle Gold Mining Co. (1931-1932). **Ore:** Gold, silver, copper. **Ore min:** Chalcopyrite, galena, sphalerite, pyrite, malachite. **Gangue:** Quartz, calcite. **Deposit:** 2- to 4-ft. vein in gneiss. **Dev:** 200-ft. adit, 260-ft. adit, 50-ft. winze, 80-ft. drift. **Assays:** 90 tons shipped prior to 1902 returned $67 per ton. **Prod:** 90 tons prior to 1902. Produced 1939-1942. **Ref:** 33, 1907, p. 674; 1908, p. 788. **63**, p. 86. **88**, pp. 37-38. **97**, 1907, 1908, 1910, 1924, 1930, 1934, 1935, 1940-1943. **106**, 7/16/31, 2/18/32, 5/19/32. **116**, no. 1, 1907, p. 17. **158**.

Highland (Highland Light) (105)

(See also Hidden Treasure.)

Loc: E½ sec. 11 and W½ sec. 12, (30-22E), Squaw Cr. dist. **Access:** 2 mi. of steep truck road from the Squaw Cr. road. **Prop.** 7 claims: Highland Light, Hidden Treasure, Buckhorn, Lookout, Sailor Boy, V. O. B., California, and millsite. **Owner:** Highland Mining & Milling Co. (1948). **Ore:** Gold, copper, tungsten, zinc. **Ore min:** Pyrite, chalcopyrite, scheelite, sphalerite. **Deposit:** Several quartz veins; one, the Sailor Boy, has an av. width of 6 ft. and is exposed for length of 3,000 ft. **Dev:** Several old adits and shafts totaling 2,000 ft. **Assays:** 2 to 5 oz. Au, 1% to 9% Zn, and very small percentage of W. **Prod:** $13,000 from 1938 to 1941 in gold and zinc. **Ref: 37,** p. 47. **63,** p. 86. **88,** p. 38. **97,** 1940, p. 478; 1941, p. 474.

Highland Light

(See Highland.)

Holden-Campbell (101)

(See also Grubstake, Hunter, Okanogan.)

Loc: Mainly secs. 10 and 11, (30-22E). Also along S. edge of secs. 2, 3, and 4, and N½ sec. 14. **Elev:** 3,000 ft. **Access:** Easily accessible from Methow highway by 4 mi. of road up Squaw and Grubstake Creeks. **Prop:** 29 claims. **Owner:** S. J. Holden and A. C. Campbell, Chelan, Wash. (1943). **Ore:** Gold, copper, tungsten. **Ore min:** Pyrite, chalcopyrite, molybdenite, scheelite. **Deposit:** Quartz veins in gneissoid diorite, among them the Hunter, the Okanogan, the Doris Barbara, the Bay Horse, the Ace of Diamonds, and the Esther veins. **Dev:** More than 1,100 ft. of underground workings and several surface cuts. **Assays:** Est. ¼% scheelite. **Prod:** Considerable gold ore. **Ref: 37,** pp. 45-47. **63,** p. 87. **158.**

Hotchkiss

(See Mazama Pride.)

Hunter (103)

(See also Holden-Campbell.)

Loc: NW¼ sec. 11, (30-22E), Squaw Cr. dist. **Prop:** 1 claim. Part of present Holden-Campbell property (1949). **Owner:** S. J. Holden and A. C. Campbell (1949). **Ore:** Gold, silver, copper, tungsten. **Ore min:** Chalcopyrite, pyrite, scheelite. **Deposit:** Quartz vein a few inches to 5 ft. wide adjacent to basalt dike in diorite. **Dev:** 660-ft. adit, 135-ft. adit, and a caved adit. Also some open cuts. **Assays:** Av. reported in 1902 as $32 Au, Ag, Cu. Also 0.4% to 0.6% WO$_3$. **Prod:** Small amount in 1940. **Ref: 63,** p. 87. **88,** p. 39. **97,** 1941, p. 474. **158.**

Imperial (Crown Point) (86)

Loc: NW¼ sec. 16, (36-20E). **Elev:** 3,700 ft. **Access:** About 5 mi. from Mazama on Goat Cr. road. **Prop:** 3 unpatented claims: Imperial, Calumet, Red Jacket. **Owner:** Mahlon McCain, Winthrop, Wash. (1946-1951). **Ore:** Gold, silver, copper. **Ore min:** Chalcopyrite, arsenopyrite, pyrite, pyrrhotite. **Deposit:** A silicified zone 20 to 25 ft. wide in quartzite is cut by a 1- to 3-ft. mineralized quartz vein. **Dev:** Inaccessible shaft 130 to 150 ft. deep, 25-ft. drift at collar of shaft, 400-ft. crosscut, 50-ft. shaft, 150-ft. drift, 2 other crosscuts, and a 10-ft. shaft. **Assays:** 1 ore shoot 100 ft. long shows from 0.16 to 0.95 oz. Au. Cu is est. at 3% to 4%. **Ref: 150,** p. 33. **158.**

Independence (97)

Loc: SW¼ sec. 29, (31-22E), on divide between McFarland Cr. and S. Fk. Gold Cr. **Elev:** 2,550 ft. **Access:** 17 mi. by road and 1 mi. by trail from Pateros. **Prop:** At least 2 claims. **Owner:** Methow Mining & Milling Co. (1943). **Ore:** Gold, copper, molybdenum. **Ore min:** Pyrite, chalcopyrite, molybdenite. **Deposit:** 3-ft. quartz vein in gneissic diorite. **Dev:** Inclined shaft, 380-ft. adit. **Assays:** $5 to $10 per ton (1902). **Prod:** 1940, 1942. **Ref: 88,** p. 40. **97,** 1941, p. 474. **158.**

Iron Cap and Snow Cap (90)

Loc: Sec. 34, (35-18E), on cliffs S. of North Lk. and in the small basin ¼ mi. W. of the lake. **Elev:** 6,500 ft. **Access:** 5 mi. by trail up North Cr. from Gilbert. **Prop:** 2 claims. **Ore:** Gold, silver, copper. **Ore min:** Pyrite, arsenopyrite, pyrrhotite, chalcopyrite, sphalerite. **Deposit:** Quartz veins in diorite near its contact with greenstone. The Iron Cap vein is about 10 ft. wide and extends from North Lk. to the top of the sheer cliffs S. of the lake, a vertical distance of 500 ft. **Dev:** Adits. **Assays:** 3 shipments to the Tacoma smelter av. 1.27 oz. Au, 1.81 oz. Ag, 0.24% Cu. **Prod:** 3 shipments to the Tacoma smelter in 1940. **Ref: 158.**

Jessie (176)

(See also Triune.)

Loc: Sec. 10, (39-26E), Wannacut Lk. dist. **Prop:** Part of Triune property. **Ore:** Gold. **Ref: 12,** p. 65. **13,** p. 103. **63,** p. 98.

John Judge

(See Leadville.)

Josie

(See Hiawatha.)

Last Chance (109)

Loc: Sec. 24, (30-22E), Squaw Cr. dist. **Ore:** Gold, silver. **Assays:** 3 carloads av. $39 per ton (1897). **Prod:** 3 carloads ore prior to 1897. **Ref: 63,** pp. 87-88.

Leadville (John Judge, Denver City, Grandview) (170)

Loc: Near SW. cor. sec. 19, (39-26E), on W. face of Palmer Mtn. **Elev:** 2,500 to 3,000 ft. **Access:** Road from Loomis. **Prop:** 13 patented claims, including: Leadville, Denver City. **Owner:** Grandview Gold Mining Co. (1941). John Judge Mining Ass'n. (1938). **Ore:** Gold, copper, lead, silver. **Ore min:** Galena, pyrite, chalcopyrite, free gold. **Gangue:** Quartz, calcite. **Deposit:** Veins of quartz 6 in. to 3 ft. wide along the contact of argillite and gabbro. Leadville vein traceable for 200 ft. in lower adit. Denver City vein is traceable for 1,000 ft. on the surface. Mineralization is sparse. **Dev:** 2,500 ft. of adits and shafts. **Prod:** 1937 (15 tons), 1938, 1939. **Ref: 63,** p. 100. **97,** 1939, p. 490; 1940, p. 478. **114,** no. 5, 1909, p. 80. **154,** pp. 104-105. **158.**

London

(See Methow.)

Mazama Pride (Hotchkiss) (89)

Loc: N½ sec. 30, (36-20E). **Elev:** 2,000 to 2,800 ft. **Access:** About ½ mi. N. of Mazama, 56 mi. by road from railroad at Pateros. **Prop:** 9 unpatented claims. **Owner:** H. A. and Edgar Hotchkiss, Winthrop, Wash., and D. W. Tomlinson, Mansfield, Wash. (1946). Danlee Mining Co. (1931). **Ore:** Gold, silver, copper. **Ore min:** Pyrite, arsenopyrite, chalcopyrite. **Deposit:** 2 or 3 quartz veins 1 to 3 ft. thick in diorite. **Dev:** 525-ft. crosscut, 80-ft. drift, 30-ft. drift, 15-ft. shaft, several open cuts. **Assays:** 37 tons shipped reportedly assayed 0.7 oz. Au. Other assays show 1.08 oz. Au, 1.50 oz. Ag. **Prod:** About 37 tons in 1931 and reportedly another shipment in 1939. **Ref: 52,** p. 11. **58,** p. 42. **97,** 1931, p. 477; 1940, p. 478. **106,** 5/7/31; 5/21/31, p. 17; 7/2/31, p. 18. **158.**

Mazama Queen (Continental) (85)

Loc: SE¼ sec. 14, (36-19E). **Elev:** 2,600 ft. **Access:** 3 mi. NW. of Mazama by road. **Prop:** 3 unpatented claims: Mazama Queen, K. D., Daisy; and 1 millsite. **Owner:** Leybold-Scales, Inc., Tacoma, Wash. (1946). Mazama Queen Mining Co. (1932). Continental Gold Silver Mining Co. (1932-1936). **Ore:** Gold, silver, copper, lead, zinc. **Ore min:** Pyrite, sphalerite, chalcopyrite, galena. **Deposit:** An 8- to 10-in. quartz-calcite vein in altered andesite. **Dev:** 1,000-ft. adit. **Improv:** 50-ton flotation plant, assay office, camp facilities for 15 to 20 men (1939). **Assays:** Ore said to range from $10 to $47 per ton. 4 samples showed 0.32 to 0.44 oz. Au, 1.74 to 4.30 oz. Ag, 8.5% to 10.0% Zn. **Prod:** Test shipments in 1938 and 1939. **Ref: 97,** 1939, p. 490; 1940, p. 478. **104,** 5/30/32, p. 26; 4/30/35, p. 29; 12/15/35, p. 26; 2/15/36, p. 31. **106,** 5/19/32; 7/7/32, p. 11. **158.**

Methow (London, New London) (106)

(See also Roosevelt.)

Loc: E½ secs. 12 and 13, (30-22E) and W½ secs. 7 and 18, (30-23E), on NE. slope of Hunter Mtn., Squaw Cr. dist. **Access:** Methow highway crosses E. end of property and is connected with W. end by truck road. **Prop:** At least 4 claims E. of Hunter Mtn. on W. side of Methow R. and at least 3 claims E. of the river. **Owner:** Methow Mining & Milling Co., Seattle, Wash. (1943-1946). Methow Gold & Copper Co. (1902-1924). **Ore:** Gold, silver, copper, tungsten. **Ore min:** Pyrite, chalcopyrite, scheelite. **Deposit:** Mineralized quartz veins from 3 in. to 15 ft. wide; among them the New London, Homestake, Mineralite, Roosevelt, Tung-stic, Milwaukee. **Dev:** More than 2,200 ft. of underground work-ings, of which more than half are W. of the river, also numerous trenches and open cuts. **Assays:** Av. 0.44 oz. Au, 0.8 oz. Ag, 1% Cu, and a small amount of W. **Prod:** $40,000 in gold in 1940-1941. **Ref: 22**, p. 8. **33**, 1907, p. 784. **37**, pp. 47-49. **63**, p. 88. **88**, p. 35. **98**, 1918-1925. **158.**

Mid Range (91)

Loc: Sec. 34, (35-18E), at head of North Cr. **Elev:** 6,000 to 7,760 ft. **Access:** About 5 mi. by trail N. of Gilbert, terminus of the Twisp R. road. **Prop:** 8 unpatented claims. **Owner:** Bill Johnson and Corvin Johnson, Winthrop, Wash. (1946). **Ore:** Gold, silver, copper, zinc. **Ore min:** Pyrite, arsenopyrite, pyrrho-tite, chalcopyrite, sphalerite. **Deposit:** 2 mineralized quartz veins in diorite. **Dev:** Several adits and open cuts. **Assays:** Ore shipped in 1940 av. 1.27 oz. Au, 1.81 oz. Ag, 0.24% Cu. **Prod:** 10 tons shipped to Tacoma smelter in 1939 and 22 tons in 1940. **Ref: 58**, p. 44. **97**, 1940, p. 478. **158.**

Minnie (95)

Loc: NW¼ sec. 23, (32-22E), Twisp dist. **Elev:** 2,400 ft. **Ac-cess:** Road up Leecher Cr. 3½ mi. NE. of Carlton. **Prop:** 4 un-patented claims. **Owner:** Franklin C. Blocksom, Twisp, Wash. (1949). **Ore:** Gold, silver, zinc. **Ore min:** Pyrite, sphalerite, chal-copyrite, scheelite, marcasite. **Gangue:** Quartz, gypsum, native sulphur. **Deposit:** Leached and honeycombed quartz vein av. 3 ft. wide in metamorphic rocks. **Dev:** 160-ft. adit on which is 55-ft. winze, 25-ft. drift, and 30-ft. stope; also several open cuts. **Improv:** House (1951). **Assays:** Carload of ore shipped in Nov. 1945: 0.46 oz. Au, 7.75 oz. Ag, net $667.56. **Prod:** 2 carloads, one in 1941, one in 1945. **Ref: 37**, pp. 52-53. **43**, 1918, p. 309. **68**, p. 14. **97**, 1918, p. 42. **158.**

Molson

(See Poland China.)

Mother Lode

(See Crystal Butte.)

Mountain Beaver (84)

Loc: NW¼SE¼ sec. 15, (38-20E), on Isabella Ridge adjacent to Billy Goat prospect. **Elev:** 4,400 to 4,800 ft. **Access:** 27 mi. by road from Winthrop. **Prop:** 3 unpatented claims: Mountain Beaver, Buckhorn Nos. 1 and 2. **Owner:** H. A. and Edgar Hotch-kiss, Winthrop, Wash., and D. W. Tomlinson, Mansfield, Wash. (1946). **Ore:** Gold, copper, silver, bismuth. **Ore min:** Pyrite, chalcopyrite. **Deposit:** Mineralized andesite agglomerate. **Dev:** 4 adits; one 275 ft. long, another with 200 ft. of work, 2 shorter ones. **Improv:** Cabin and small flotation mill (1949). **Assays:** Crude ore shipped av. 1.53 oz. Au, 1.13 oz. Ag, 1.81% Cu. Minor amount of bismuth reported. **Prod:** Small shipments in 1922. 1931, 1934, 1935. **Ref: 97,** 1931, p. 477; 1935, p. 354. **158.**

New London

(See Methow.)

Northern Gold

(See Bodie.)

Occident (179)

(See also Triune.)

Loc: Sec. 10, (39-26E), Wannacut Lk. dist. **Prop:** Part of Triune property. **Ore:** Gold, silver. **Ref: 12,** p. 65. **13,** p. 104.

Okanogan (102)

(See also Holden-Campbell.)

Loc: W. of Hunter adit, secs. 10 and 11, (30-22E), Squaw Cr. dist. **Prop:** 1 claim of Holden-Campbell group. **Owner:** Holden-Campbell Mining Co. (1942). **Ore:** Gold, tungsten. **Ore min:** Scheelite. **Deposit:** Quartz vein as much as 8 ft. wide in granodiorite. Vein exposed on surface for at least 300-ft. length and 142-ft. depth. **Dev:** 160-ft. adit (caved 1942). **Assays:** Est. $\frac{1}{4}\%$ scheelite. **Ref: 37,** pp. 45-47. **158.**

Okanogan Free Gold (Owasco, Allison) (175)

Loc: SE¼NW¼ sec. 19, (40-27E), on N. side of Similkameen R. **Elev:** 1,200 ft. **Access:** 3 mi. N. of Oroville by road. **Prop:** 5 patented claims. **Owner:** Roy S. Meader, Oroville, Wash., and H. N. North, Omak, Wash. (1938-1941). United Mines Co. (1914). Owasco Gold Mining Co. (1915-1918). **Ore:** Gold, silver. **Ore min:** Pyrite, sphalerite, galena, free gold. **Deposit:** A quartz vein 0 to 12 ft. wide occurs in country rock of limestone, quartzite, and schist. Ore minerals are disseminated in the quartz. **Dev:** 3 adits, glory hole. **Assays:** 2 assays show $3.92 and $5.37 Au, Ag. **Prod:** 1914, 1918, 1936, 1938-1939. **Ref: 54,** pp. 4-5. **58,** pp. 3, 52. **63,** p. 102. **97,** 1914, 1918, 1937, 1939-1940. **100,** 1899, p. 80. **112,** p. 196. **154,** p. 87. **158.**

Oriental and Central

(See American Flag.)

Overtop

(See Poland China.)

Owasco

(See Okanogan Free Gold.)

Palmer Mountain Tunnel (173)

Loc: NE¼NE¼ sec. 1, (38-25E), Palmer Mtn. dist., ½ mi. N. of Loomis. **Elev:** 1,610 ft. **Prop:** 56 unpatented claims. **Owner:** Tillicum Development Co. (1912-1924). Palmer Mountain Gold Mining & Tunnel Co. (1902-1918). Palmer Mountain Tunnel Co. (1903-1918). Palmer Mountain Tunnel & Power Co. (1905-1908). Palmer Mountain Gold & Copper Co. (1908). Palmer Mountain Mining & Milling Co. 1908-1918). **Ore:** Gold, silver, copper, lead. **Ore min:** Malachite, pyrite, chalcopyrite, galena. **Deposit:** Quartz veins as much as 3 ft. wide in diorite and greenstone. Smaller veins contain much calcite. **Dev:** 6,610-ft. adit, and drifts aggregating 2,000 ft. An 800-ft. diamond drill hole extends beyond face of main adit. **Assays:** 20-ft. sample from 3-ft. vein in tunnel 1,300 ft. from portal assayed 0.56 oz. Au, 0.28 oz. Ag. Other assays from $1.80 to $40 per ton. **Ref:** 33, 1908, p. 1099. **88,** pp. 29-30. **98,** 1918-1925. **100,** 1903, p. 56. **104,** 3/30/34, p. 25. **145,** p. 95. **154,** pp. 106-107. **158.**

Palmer Summit (Grand Summit) (182)

Loc: SE¼SE¼ sec. 21, (39-26E), Palmer Mtn. dist. **Owner:** Dr. L. E. Eastman and F. C. Barber, Wenatchee, Wash. (1939-1941). **Ore:** Gold, copper, lead. **Ore min:** Pyrite, chalcopyrite, galena. **Deposit:** Narrow and sparsely mineralized quartz veins in gabbro. **Dev:** 50-ft. adit, a second adit, and a 40-ft. shaft. **Assays:** 50 tons av. $20 Au. **Prod:** $1,000 prior to 1897. 1937, 1939. **Ref:** 58, p. 52. **63,** p. 100. **97,** 1938, p. 459; 1940, p. 478. **158.**

Pateros

(See Sullivan.)

Paymaster (110)

Loc: SW¼NW¼ sec. 14, (30-22E), Squaw Cr. dist. **Owner:** S. J. Holden and A. C. Campbell, Chelan, Wash. (1942). **Ore:** Gold. **Ore min:** Free gold, scheelite. **Deposit:** A vein 3 ft. wide composed of sheared granitic rock and quartz. Iron oxide bands from 1 in. to 3 in. wide on each wall of the vein carry free gold. **Dev:** 150-ft. shaft, 175-ft. adit. **Prod:** Some gold ore produced from the oxidized ore. **Ref:** 58, p. 53. **63,** p. 88. **158.**

Pennington

(See Chelan.)

Pinnacle (169)

Loc: SW¼ sec. 19, (39-26E) and sec. 25, (39-25E). **Elev:** 2,000 ft. **Access:** 3 mi. N. of Loomis on Palmer Lk. road. **Prop:** 6 claims. **Owner:** Pinnacle Mines, Inc., Renton, Wash. (1951). Pinnacle Gold Mine Co. (1909-1915). Silver-Lead Mining & Reduction Co. (1929). Sinlahekin Mining Co. (1931). Mary M. Mining Co.

(1935). **Ore:** Gold, copper, lead, zinc, silver. **Ore min:** Free gold, pyrite, chalcopyrite, sphalerite. **Gangue:** Quartz, some calcite. **Deposit:** Quartz vein 4 to 10 ft. wide in altered gabbro, and disseminated ore minerals in sheared gabbro. **Dev:** Approx. 2,000 ft. of workings in 3 adits. **Assays:** $11 Au across the vein 4 to 10 ft. wide. **Prod:** $200,000 prior to 1910 (gold). **Ref:** **99,** 3/26/35. **106,** 8/2/28, 10/1/31. **114,** no. 5, 1909, p. 80. **150,** p. 38. **154,** p. 105. **158.**

Poland China (Molson, Overtop) (187)

Loc: SW¼SE¼ sec. 11, (40-29E), Myers Cr. dist. **Elev:** 3,850 ft. **Access:** About 4 mi. NW. of Chesaw. 10 mi. to railroad at Mancaster, B. C. **Prop:** 11 patented claims, including the Charles F. and Kismet. Also 160 acres on which deed to mineral rights is held. **Owner:** Overtop Mining Corp., Molson, Wash. (1932-1940). Molson Mining Co. (1907-1918). Mary Ann Creek Mining Co. (1922). **Ore:** Gold, silver, lead. **Ore min:** Pyrite, marcasite. galena. **Deposit:** Quartz vein from a stringer to 14 ft. wide enclosed in graphitic argillite. Av. width 6 ft. **Dev:** 1,500 ft., consisting of a 436-ft. adit, various shafts, and crosscuts. **Assays:** Ore worked in 1911 av. about $4.75 Au, 1 oz. Ag. 2 cars shipped in 1938 assayed $13 Au. **Prod:** More than $100,000 by 1936, 1937, at least 2 cars in 1938, 1939. **Ref:** **63,** p. 110. **97,** 1907, 1913, 1914, 1923, 1934, 1935, 1938-1940. **98,** 1925, p. 1823. **104,** 7/30/36, p. 25. **106,** 7/30/32; 5/19/32; 7/21/32, p. 1; 7/30/32, pp. 28-29. **112,** p. 192. **113,** 4/15/37, p 7. **116,** no. 6, 1907, p. 33. **154,** pp. 50-51. **158.**

Rainbow (183)

Loc: NE¼ sec. 22, (39-26E), between Palmer Mtn. and Wannacut Lk. **Elev:** 2,600 ft. **Access:** Road from Loomis or Golden. **Prop:** 5 claims: Rainbow, Mayflower, Coyote, Cottonwood, June Bug. **Owner:** Wannacut-Rainbow Mining Co. (1930). Washington Mining & Development Co. (1892). **Ore:** Gold, silver, copper. lead. **Ore min:** Pyrite, arsenopyrite, chalcopyrite, galena, malachite, limonite, free gold. **Deposit:** Vein consists of quartz lenses enclosed in limestone, quartzite, and schist. The lenses pinch and swell within short distances. **Dev:** 3 adits. **Assays:** Rich free gold samples assayed 50 to 60 oz. Au, 18 to 20 oz. Ag, 2% Cu. 2.9% to 3.5% Pb. **Prod:** Has produced. **Ref:** **12,** p. 66. **13,** p. 101. **54,** p. 3. **63,** p. 100. **106,** 10/2/30. **154,** p. 102.

Reco (190)

Loc: SW¼ sec. 16, (40-30E), ½ mi. N. of Chesaw, Myers Cr. dist. **Access:** Road. **Prop:** 2 claims: Reco, Pendennis. **Owner:** Reco Gold Mining Co., Bremerton, Wash. (1915-1938). Reco Mining & Milling Co. (1918). **Ore:** Gold, silver, copper. **Ore min:** Arsenopyrite, free gold, chalcopyrite, pyrite, marcasite, bornite.

Gangue: Quartz, some calcite. **Deposit:** Veins 1 ft. to 2 ft. wide in granitic rock. **Dev:** About 1,100 ft. in 3 adits, also some open cuts. **Assays:** Ore shipped said to av. $20 per ton. **Prod:** 150 tons shipped to Northport smelter in 1916 and 1917. **Ref: 58,** p. 57. **97,** 1914-1916, 1924. **106,** 12/1/32. **112,** p. 200. **158.**

Red Shirt (93)

Loc: SE¼ sec. 18, (33-23E), on lower W. slope of Pole Pick Hill. **Elev:** 3,800 ft. **Access:** Road from Twisp. **Prop:** 1 patented claim: Red Shirt; 4 unpatented claims. **Owner:** John Russell and George M. Gibson, Winthrop, Wash., lessees (1952). Red Shirt Mining Co. (1936-1938). Mahlon McCain, Winthrop, Wash. (1949). **Ore:** Gold, silver, copper. **Ore min:** Pyrite, chalcopyrite, arsenopyrite. **Deposit:** Quartz vein, with a width of 1 ft. to 5 ft., in schist. **Dev:** 100 ft. of drift from a 425-ft. crosscut, and 375 ft. of drift from a 200-ft. crosscut. **Assays:** 35 oz. Ag, ½ oz. Au. Ore av. about $10 per ton. **Prod:** Credited with more than $100,000. Produced intermittently for 50 yr. Latest work in 1936 to 1938. **Ref: 12,** p. 69. **13,** p. 113. **63,** p. 85. **97,** 1937-1939. **104,** 9/30/36, p. 30. **158.**

Roosevelt (108)

(See also Methow.)

Loc: SW¼NW¼ sec. 18, (30-23E), ¼ mi. E. of the Methow R., Squaw Cr. dist. **Elev:** 75 ft. above Methow R. **Owner:** Methow Mining & Smelting Co. (1930-1945). **Ore:** Gold, copper, tungsten. **Ore min:** Pyrite, chalcopyrite, scheelite. **Deposit:** Quartz vein associated with basaltic dike in gneissic granitic rock. It varies from less than 1 in. to 6 ft. in thickness and av. 2½ to 3 ft. Sulfides occur in pods on walls of the vein. Scheelite is in pockets and stringers throughout the vein. The vein is exposed in adits and on the surface for length of more than 700 ft. and depth of 400 ft. **Dev:** 665-ft. drift, a lower 240-ft. drift, both caved in 1938. **Assays:** Less than ¼% scheelite. Assays for the back 180 ft. of the lower adit av. 0.50% to 0.75% WO$_3$ over an av. width of 2 ft. Lower scheelite values in the upper adit. **Ref: 37,** p. 49. **40,** p. 15. **97,** 1929, p. 426; 1930, p. 673. **106,** 4/3/30. **158.**

St. Anthony (99)

Loc: SW¼SW¼ sec. 5, (30-22E), Squaw Cr. dist. **Access:** McFarland Cr. road, 12 mi. to railroad. **Prop:** 9 unpatented claims. **Owner:** Holden Gold Mines, Inc., Chelan, Wash. (1934). Squaw Creek Mining Co. (1931). **Ore:** Gold, silver, copper. **Ore min:** Free gold. **Gangue:** Quartz. **Deposit:** Vein in gneissic diorite is 5 ft. wide. **Dev:** Adit consisting of 90 ft. of crosscut and 125 ft. of drift, a shaft of unknown depth, and several open cuts. **Assays:** Ore from surface to 35-ft. depth said to run $50 Au.

Prod: 1934. **Ref:** **58**, p. 64. **98**, 1935, p. 354. **104**, 7/15/33, p. 19; 12/30/33, p. 19. **106**, 4/16/31. **158**.

Silver Bell (196)

Loc: Secs. 25 and 36, (38-31E) and secs. 30 and 31, (38-32E), 12 mi. NW. of Republic. **Prop:** 6 claims: Silver Bell, Silver Bell No. 2, Uncle Sam No. 1, Uncle Sam Extension, Valley, Valley Extension. **Owner:** New Silver Bell Mining Co., Inc., Almira, Wash. (1949). **Ore:** Gold, silver. **Deposit:** 2 ore shoots reported. **Dev:** 340-ft. adit. **Assays:** $3.28 to $28 per ton reported from one ore shoot. **Prod:** High-grade ore reported shipped from surface pit prior to 1907. In 1940 a 28-ton shipment gave net smelter returns of $244.03. **Ref:** **58**. **68**, p. 14. **158**.

Silver Ledge (96)

Loc: Near center sec. 11, (31-21E), ¼ mi. downstream from Antimony Queen mine. **Elev:** 300 ft. above Gold Cr. **Access:** 4 mi. by road up Gold Cr. from Methow highway. **Prop:** 3 claims: North Star, Truax, Seattle. **Owner:** Silver Ledge Mining Co. (1920-1931). **Ore:** Gold, silver. **Ore min:** Pyrite, arsenopyrite, silver sulfide, cerargyrite, bromyrite. **Deposit:** Quartz veins in sandstone and shale. Vein at collar of shaft is in a 5-ft. shear zone and consists of 2 parts, one 1½ ft. wide at footwall and the other 4 in. wide on hanging wall. **Dev:** 150-ft. inclined shaft, 1,500-ft. crosscut. **Assays:** 2 tons shipped av. $20 per ton. **Prod:** 2 tons of unsorted ore prior to 1921. **Ref:** **98**, 1922-1926. **104**, 12/15/31, p. 29. **113**, 7/34. **129**, p. 259. **158**.

Snow Cap

(See Iron Cap and Snow Cap.)

Spokane (American Rand) (180)

Loc: Near center E½ sec. 10, (39-26E), above NW. shore of Wannacut Lk. **Access:** Good road from Oroville. **Prop:** 13 patented and 3 unpatented claims. **Owner:** American Rand Corp. (1935-1938). Morgan Mines Co. (1916-1918). Oroville Gold Mining Co. (1934). **Ore:** Gold, silver, lead, copper, molybdenum. **Ore min:** Pyrite, galena, chalcopyrite, molybdenite, reportedly free gold. **Deposit:** Quartz vein 15 in. wide in argillite a short distance above a granite contact. Ore minerals occur in fractures in the quartz. Pyrite disseminated in the wall rock. **Dev:** 100-ft. inclined shaft from which short drifts have been driven. **Assays:** Ore from adit: 8 oz. Au, 2 oz. Ag. From shaft: 7.5 oz. Au, 2 oz. Ag. From dump: 7.5 oz. Au, 2.5 oz. Ag. 1.05% molybdenum reported. **Prod:** 1916-1918, 1935-1938. **Ref:** **12**, p. 66. **13**, p. 104. **63**, pp. 99-100. **97**, 1916-1918, 1935, 1937, 1939. **104**, 2/15/35, p. 23; 6/30/36, p. 25. **154**, p. 98. **158**.

Spokane (Gold Crown) (92)

Loc: Sec. 12, (33-21E). **Access:** On Twisp R. road, 2 mi. W. of Twisp. **Prop:** 2 unpatented claims. **Owner:** S. J. Sherwood, Twisp, Wash. (1946). **Ore:** Gold, silver, lead, zinc, copper. **Ore min:** Sphalerite, arsenopyrite, chalcopyrite, pyrite, galena. **Deposit:** Irregular quartz-calcite vein from few inches to 3 ft. thick in andesite. Ore minerals occur in bunches in vein. **Dev:** Several hundred ft. of crosscuts and drifts and some stopes. **Assays:** 6 tons shipped in 1941 assayed 0.40 oz. Au, 20.25 oz. Ag, 4.5% Zn, 3.8% Pb. **Prod:** Small amount in 1939 and 1941. **Ref:** 1, no. 2, 1917, p. 29. **58**, pp. 24, 64. **88**, p. 36. **97**, 1940, p. 478. **105**, no. 3, 1906, p. 46. **158**.

Sullivan (Pateros)

Loc: 5 mi. NW. of Pateros, on E. side of Methow R. **Access:** Mine is across Methow R. from highway, by which it is 5 mi. to railroad at Pateros. **Prop:** 5 claims: Friday, Saturday, Key, Lone Pine, Prince of Paris. **Owner:** J. J. Sullivan, Pateros, Wash. (1947), leasing to Dr. Roy W. Key (1940-1947). Pateros Mining Co. (1943). **Ore:** Gold, silver, copper. **Ore min:** Pyrite, chalcopyrite. **Gangue:** Quartz, carbonate minerals. **Deposit:** Sheared zones up to 6 ft. thick but av. less than 3 ft. in granite along margins of "andesite" dikes contain small scattered pods of shipping-grade ore. **Dev:** 1,200 ft. of underground workings on 3 levels. A block of ore 75 ft. long, 30 ft. high, and up to 6 ft. thick was stoped above the upper adit in early days. **Assays:** $6 to $30 per ton, mostly in gold. 7 tons shipped in 1940 showed 0.63 oz. Au, 0.80 oz. Ag, 0.17% Cu. **Prod:** Reported 60 carloads of ore prior to 1897 valued at $72,000. 1940, 1941. **Ref:** 157. 158.

Summit (172)

(See also Alice.)

Loc: Sec. 30, (39-26E), on summit of Palmer Mtn. **Access:** 12 mi. by road to railroad at Tonasket. **Prop:** 1 patented claim: Summit; 7 unpatented claims: Edgeworth, Latakin, Copenhagen, Nicotine, Velvet, Half-and-Half, Tuxedo. **Owner:** R. D. Heffernan, Tonasket, Wash. (1949). Grand Summit Mining Co. (1936-1939). **Ore:** Gold, silver. **Ore min:** Free gold. **Gangue:** Quartz. **Deposit:** 2 veins. **Dev:** 3 shafts totaling 420 ft., 5 adits totaling 1,200 ft. **Assays:** Av. about $20. Mill recovered $1,538 from 85 tons in 1937. **Prod:** 1937 ($1,900), 1938. **Ref:** 52, p. 5. 104, 11/15/36, p. 27. 113, 11/25/36, p. 18. 158.

Sunshine

(See Hidden Treasure.)

Tom Hal

(See Friday.)

Triangle

(See Hidden Treasure.)

Triune (Crescent) (181)

(See also Jessie, Occident.)

Loc: NE¼ sec. 10, (39-26E), ½ mi. NW. of the Kimberly property, Wannacut Lk. dist., ½ mi. W. of the old post office at Golden. Elev: 2,300 ft. Prop: 5 claims, including Triune, Jessie, Occident. Owner: Dell Hart, Oroville, Wash. (1947-1954). Washington Consolidated Mining Co. (1905). Triune Power & Reduction Co. (1911). Dell Hart & Sons Mining Co. (1938). G. O. F. Triune Mining Co. (1939). Ore: Gold, silver, lead, copper. Ore min: Pyrite, galena, chalcopyrite, molybdenite, free gold, malachite, azurite. Deposit: At least 4 quartz veins varying in width from a stringer to 10 ft. occur in argillite a short distance above intrusive granite. The granite is sericitized and kaolinized. Dev: 140-ft. shaft with adit on lower level together with drifts total more than 2,000 ft. Assays: Early assays $10 to $40 per ton. An av. assay of 4,000 tons of tailings gave 0.01 oz. Au, 0.3 oz. Ag. Sample of 3 ledges on S. side of gulch (veins from 4 to 10 ft. wide) assayed 0.32 oz. Au, 0.5 oz. Ag. Prod: More than $300,000 prior to 1938, 1939. Ref: 12, p. 65. 13, p. 104. 51, p. 11. 63, p. 99. 88, p. 31. 98, 1905, p. 336; 1934, p. 297; 1939, p. 490. 106, 11/3/32, p. 9. 108, 10/39, p. 31. 154, pp. 97-98. 158.

War Eagle (185)

(See also Alice.)

Loc: SE¼ sec. 36, (39-25E), on Palmer Mtn. Prop: 1 claim. Owner: R. D. Heffernan, Tonasket, Wash. (1949). Ore: Gold, copper, silver. Ore min: Free gold. Deposit: 5-ft. quartz vein. Dev: 100-ft. adit, two 70-ft. drifts. Assays: Ore av. about $18 Au, Cu, Ag. Values chiefly in Au. Prod: Has produced. Ref: 12, p. 63. 13, p. 99. 52, p. 5. 63, p. 98. 88, p. 31.

Whitestone (194)

Loc: Sec. 34, (39-30E), on S. slope of Strawberry Mtn., Wauconda dist. Owner: Whitestone Mining Co. (1938). Ore: Gold, silver, lead, zinc, antimony. Ore min: Galena, sphalerite, tetrahedrite. Deposit: Limestone beds with low dips are mineralized on tops and bottoms and along stringers cutting the beds. Prod: 1938. Produced intermittently 1918-1938. Ref: 58, p. 73. 97, 1939, p. 490. 158.

PEND OREILLE COUNTY

Deemer (253)

Loc: Sec. 2, (39-45E), Metaline dist. Access: Road. Prop: 5 unpatented claims. Owner: Jack Doyle, Metaline Falls, Wash. (1941). Ore: Gold, silver. Deposit: Quartz vein from a few in. to

4 ft. in width is said to carry gold and silver. **Dev:** 200 ft. of adits and numerous open cuts, mostly caved. **Ref: 29,** p. 38.

Gilbert (254)

Loc: Sec. 24, (31-44E), Newport dist. **Prop:** 160 acres of deeded land. **Owner:** Gilbert Fellen, Newport, Wash. (1941). **Ore:** Gold. **Deposit:** Said to be a gold-bearing quartz vein 6 in. wide which has been traced for 200 ft. on the surface. **Dev:** Caved open cuts. **Ref: 29,** p. 73.

Hansen (256)

Loc: Sec. 14, (30-43E), Newport dist. **Access:** 1 mi. from road. **Prop:** 160 acres of deeded land. **Owner:** Mr. Hansen, Elk, Wash. (1941). **Ore:** Gold, lead, zinc, copper. **Deposit:** Fracture zone said to be impregnated with narrow mineralized quartz stringers. Zone is 20 ft. wide. **Dev:** Inclined shaft, caved. **Ref: 29,** pp. 74-75.

Rocky Creek (252)

Loc: SE. cor. sec. 23, (37-41E). **Access:** About 12 mi. by road from railroad at Tiger. **Prop:** 3 claims. **Owner:** Kaniksu Metals, Tacoma, Wash. (1954). George and Henry Rushmier and Mrs. Bellows, Colville, Wash. (1953). **Ore:** Silver, lead, zinc, gold. **Ore min:** Galena, chalcopyrite, sphalerite, free gold. **Deposits:** Fine-grained sulfides in a quartz vein 2 to 16 in. wide exposed for 50 ft. in a schist roof pendant in granite. **Dev:** 10-ft. shaft, 70-ft. drift. **Assays:** Av. 360 oz. Ag. One assay showed 0.44 oz. Au. Some ore carried free gold to the extent of several thousand dollars per ton. **Ref: 158.**

Sunrise (255)

Loc: Sec. 22, (30-44E), Newport dist. **Access:** Road. **Prop:** 160 acres of deeded land. **Owner:** C. A. Vanderholm, Camden, Wash. (1941). **Ore:** Gold, silver. **Deposit:** Quartz vein from a few in. to 3 ft. in width is said to carry values in gold and silver. **Dev:** 250 ft. of adit and shaft, caved (1941). **Improv:** Several buildings (1941). **Assays:** Handpicked sample assayed $10.50 per ton. **Ref: 29,** p. 72.

PIERCE COUNTY

Bear Gap
(See Fife under Yakima County.)

Blue Bell
(See Fife under Yakima County.)

Blue Grouse and Sure Thing (65)

Loc: NE¼ sec. 36, (17-10E), Summit dist. **Ore:** Gold, silver. **Deposit:** Series of parallel veins. **Assays:** $3 to $51 Au, 31 oz. Ag. **Ref: 63,** p. 45.

Campbell (66)

Loc: Sec. 36, (17-10E), on S. end of Crystal Mtn. at the head of Silver Cr. **Elev:** 5,500 to 6,000 ft. **Access:** 3 mi. by trail from end of Morse Cr. road. **Prop:** 5 unpatented claims. **Ore:** ·Gold. **Ore min:** Free gold. **Deposit:** Small quartz veins in volcanic rock. Both veins and volcanics reputedly carry gold values. **Assays:** Gold values from 20¢ to a few dollars per ton. **Ref:** 158.

Fife

(See under Yakima County.)

New Deal

(See Washington Cascade.)

Pickhandle

(See Fife under Yakima County.)

Seigmund Ranch (63)

Loc: Sec. 30, (17-5E), 1 mi. from Clay City. **Owner:** E. J. Cowan, Tacoma, Wash. (1952 ——). Seigmund Gold Mining Co. (1920). **Ore:** Gold. **Ore min:** Free gold. **Deposit:** Vein 20 to 40 ft. wide of free-milling gold-bearing quartz. **Assays:** Gold values are very low. **Ref: 69-A,** p. 15. **130,** pp. 112-113. **158.**

Silver Creek (67)

Loc: Sec. 25, (17-10E), at headwaters of Silver Cr. and its tributaries, Summit dist. **Access:** Trail from end of road at head of Silver Cr. **Owner:** J. B. Garrard and H. H. Lentz (1945). **Ore:** Gold, silver, copper, zinc, lead. **Ore min:** Pyrite, chalcopyrite, pyrrhotite, some marcasite, sphalerite, and arsenopyrite. **Gangue:** Quartz, calcite. **Deposit:** Altered and silcified andesite is mineralized along narrow joints and in places heavily impregnated with pyrite. **Dev:** 42-ft. shaft, 28-ft. drift, 300-ft. crosscut, 25-ft. adit, 38-ft. adit, 70-ft. adit, several open cuts. **Assays:** 100 tons shipped reportedly ran more than $50 per ton. **Prod:** 100 tons of ore prior to 1945; 3 oz. Au, 14 oz. Ag, 100 lb. Pb in 1941. **Ref: 97,** 1941, pp. 484, 487. **104,** 8/15/34, p. 23. **158.**

Silver Creek Gold & Lead (64)

Loc: Sec. 12, (17-10E), Summit dist. **Access:** 35 mi. by road to railroad at Enumclaw. 60 mi. by road to copper smelter at Tacoma. **Prop:** 6 unpatented claims and a millsite. **Owner:** Silver Creek Gold & Lead Corp., Seattle, Wash. (1951 ——). **Ore:** Gold, silver, lead, zinc, copper. **Ore min:** Arsenopyrite, pyrite, chalcopyrite, galena, sphalerite. **Deposit:** Rhyolite cut by fault zone 1 to 4 ft. wide containing limy gouge and bands of quartz with sulfides. **Dev:** 100-ft. crosscut with 100-ft. drift. **Improv:** Small cabin, rails, mine car, 10-ton ore bin (1951). **Prod:** 20 tons (1945) reported to have contained $89.75 per ton in gold, silver, lead, copper. Unknown amount of placer gold has been produced from stream gravels. **Ref: 133**, p. 39. **157.**

Sure Thing

(See Blue Grouse and Sure Thing.)

Tacoma

(See Washington Cascade.)

Washington Cascade (New Deal, Tacoma) (68)

Loc: Sec. 25, (17-10E), on Silver Cr., Summit dist. **Access:** 7-mi. road from Chinook Pass highway. **Prop:** 2 claims: New Deal, Tacoma. **Owner:** Blue Bell Gold Mining Co., Tacoma, Wash. (1940). **Ore:** Gold, silver, copper. **Deposit:** 5-ft. vein in andesite. **Dev:** 755-ft. adit. **Assays:** 30 tons returned $180. **Prod:** 1936 (5 tons), 1938 (30 tons). **Ref: 158.**

SKAGIT COUNTY

Devils Mountain

(See Mount Vernon.)

Mount Vernon (Devils Mountain, Pacific) (27)

Loc: S½ sec. 4, NE¼ sec. 9, N½ sec. 10, and NW¼ sec. 11, (33-4E), on Devils Mtn. **Elev:** 250 to 1,750 ft. **Access:** 4½ mi. SE. of Mount Vernon by road. 1½ mi. by road from railroad. **Prop:** 2,100 acres. **Owner:** Pacific Nickel Co. (1941-1943). **Ore:** Nickel, gold. **Ore min:** Chromite, free gold, nickeliferous ankerite, marcasite, pyrite, bravoite. **Gangue:** Quartz, chalcedony, carbonates. **Deposit:** Fault zone between serpentine and sandstone is made up of silica-carbonate rock with a central core of sulfide-bearing breccia. Silica-carbonate rock portion is 2 mi. long and 100 to 400 ft. wide. Sulfide breccia is in small lenses along the core. **Dev:** 300 ft. of adits and 6,375 ft. of diamond drilling. **Assays:** 157 assays on 2,598 ft. of core av. 0.0195 oz. Au, 0.251% Ni. **Prod:** Test shipments only. **Ref: 60. 81. 158.**

Pacific

(See Mount Vernon.)

Soldier Boy (26)

Loc: SE¼ sec. 25 and NW¼ sec. 36, (35-13E), SE. of Midas group. **Access:** Trail. **Prop:** 5 patented claims: Soldier Boy, Gold Run, Golden Eagle, Mountain Chief, North Star. **Owner:** A. G. Mosier, Sedro Woolley, Wash. (1951). **Ore:** Gold, silver, lead, zinc, copper. **Ore min:** Pyrite, chalcopyrite, sphalerite. **Gangue:** Quartz. **Deposit:** Said to be several veins 1 to 3 ft. wide. **Dev:** Adit. **Assays:** $27 to $100 per ton in all values reported. **Ref: 63,** p. 55. **158.**

SKAMANIA COUNTY

Brown and Livingston (79)

Loc: On Granite Cr., a tributary of McCoy Cr., 3 mi. above the mouth of McCoy Cr., probably in sec. 15, (10-8E). **Ore:** Gold, silver. **Assays:** $6.00 to $30.00 Au, Ag. **Ref: 158.**

Bruhn (74)

Loc: Sec. 10, (10-8E), adjacent to property of Camp Creek Metals Mining Co., McCoy Cr. dist. **Ore:** Gold. **Ore min:** Sulfides. **Ref: 158.**

Camp Creek (75)

Loc: Center sec. 10, (10-8E), on Camp Cr., a tributary to McCoy Cr. **Owner:** Camp Creek Metals Mining Co. (1934). **Ore:** Gold. **Ore min:** Free gold. **Deposit:** Quartz stringers in an oxidized zone 300 ft. wide. **Prod:** $75,000 prior to 1934. **Ref: 58**, p. 12. **97**, 1935, p. 354. **158.**

Golconda (72)

Loc: Near W. ¼ cor. sec. 16, (10-6E), St. Helens dist. **Access:** ½ mi. up the Ryan Lk. trail from the Green R. trail. **Ore:** Gold, zinc, copper. **Ore min:** Pyrite, some chalcopyrite and sphalerite, and a little galena. **Deposit:** Quartz veinlets along a fracture zone in granite. **Dev:** Adit. **Assays:** Reportedly $4.00 Au. **Prod:** An 18-ton shipment in 1933 reported. **Ref: 63**, p. 49. **158.**

Grizzly Creek (73)

Loc: Near Grizzly Cr. on Black Mtn., probably in sec. 20, (10-6E). **Access:** Trail. **Prop:** 4 claims. **Ore:** Gold, silver, lead, copper. **Deposit:** Vein av. 4 ft. wide in granite. **Dev:** 2 adits and a winze. **Assays:** $50.00 to $60.00 per ton reported from bottom of winze. **Ref: 158.**

Johnson (76)

Loc: Sec. 10, (10-8E), a short distance above the Bruhn property on McCoy Cr. **Ore:** Gold. **Ore min:** Free gold. **Deposit:** Gold in a seam 2 or 3 in. wide. **Ref: 158.**

Livingston

(See Brown and Livingston.)

Perry (77)

Loc: Above the Johnson property cn McCoy Cr., probably in sec. 10, (10-8E). **Ore:** Gold. **Ore min:** Arsenopyrite. **Ref: 158.**

Plamondon (80)

Loc: SW¼SE¼ sec. 7, (8-8E), on N. side of Quartz Cr. **Elev:** 2,800 ft. **Access:** 3½ mi. by trail from end of road. **Prop:** 26 unpatented claims (1943). 3 unpatented claims (1954). **Owner:** Plamondon Brothers Mining Co. (1939 ——). Plamondon Indian Mines, Inc. (1940-1944). **Ore:** Gold, silver. **Gangue:** Chalcedonic quartz. **Deposit:** Chalcedonic quartz in iron-stained tuff is reportedly mineralized. **Dev:** Short adit, open cut. **Assays:** As high as $68 reported. **Ref: 158.**

Primary Gold (78)

Loc: Sec. 10, (10-8E), on both sides of Camp Cr. **Access:** 4 mi. by trail from the end of Niggerhead road. **Prop:** 17 unpatented claims: Jim Nos. 1 to 6; Inez Nos. 1 to 4; Eva Nos. 1, 2, and 5; Jim No. 8; Ialine; O Bill; Orell. **Owner:** Primary Gold Co., Yakima, Wash. (1938). **Ore:** Gold, silver, and platinum reported. **Deposit:** 30-ft. ledge of coarsely crystalline pyrite in quartz gangue. This carries only traces of gold, but narrow stringers of free gold in quartz cut the pyrite. **Dev:** Numerous open cuts, some placering. **Assays:** Said to range from $2.10 to $1,500.00 per ton. **Prod:** Has produced (probably small amount). **Ref: 58,** p. 55. **158.**

SNOHOMISH COUNTY

Ben Butler (43)

Loc: NE¼ sec. 6, (28-11E), on W. side of Silver Cr. **Owner:** David Fergerson. **Ore:** Gold. **Deposit:** 12- to 15-ft. vein carrying 12 to 30 in. of ore. **Dev:** 60-ft. adit. **Ref: 14,** p. 29. **63,** p. 32.

Ben Lomond

(See Rainy.)

Blue Bird (28)

Loc: Sec. 24, (32-9E) and sec. 19, (32-10E), Darrington dist. **Prop:** Several claims. **Owner:** Pacific Nickel Co., Mount Vernon, Wash., leasing from Earl Scott, Bellingham, Wash. (1943-1944). Sauk River Mining Co. (1907). Albert Hawkinson (1908). **Ore:** Gold, copper, silver, nickel. **Ore min:** Chalcopyrite, pyrrhotite. **Deposit:** Vein av. 8 ft. in width. **Dev:** 200 to 300 ft. of adits. **Assays:** High-grade ore runs $30 per ton. **Prod:** Small amount reported about 1940. **Ref: 14,** p. 8. **22,** p. 9. **33,** 1907, p. 1011. **49,** p. 10. **88,** p. 62. **159,** p. 134.

Bonanza

(See Mineral Center.)

Calumet (48)

Loc: Near center N½ sec. 27, (29-10E), in Sultan Basin. **Elev:** 4,600 to 5,000 ft. **Access:** About ½ mi. from road. 26 mi. to railroad at Sultan. **Prop:** Calumet claim. **Owner:** Sultan Basin Mining Co., Robert Curtis, lessee, Monroe, Wash. (1951). **Ore:** Gold, silver, copper, zinc. **Ore min:** Chalcopyrite, pyrrhotite, marcasite, arsenopyrite, sphalerite, galena. **Deposit:** Quartz vein from 6 to 24 in. wide along a fracture zone in metamorphic rocks. **Dev:** 250-ft. adit, 9 open cuts. **Assays:** Tr. to 1.03 oz. Au, 0.56 to 5.33 oz. Ag, 0.48% to 6.50% Cu, 0.10% to 4.85% Zn. **Ref: 23**, pp. 62, 65-68. **37**, p. 56. **150**, p. 30.

Caplin-Holbrooke (38)

Loc: Sec. 33, (29-11E), on the E. Fk. of Silver Cr. about 2 mi. N. of Mineral City. **Prop:** 9 claims: Gold Eagle, New Strike, Last Chance, Mt. Beauty, Fortunate Monarch, Silver Queen, Good Grub, Cougar, Two Hams. **Ore:** Gold, silver, copper. **Deposit:** 10-ft. vein with two 12-in. paystreaks of ore. **Dev:** 75-ft. adit. **Assays:** As high as $90 Au, Ag, Cu. **Ref: 13**, pp. 159-160. **14**, p. 29.

Clara Thompson

(See Jasperson.)

Commonwealth

(See Jasperson.)

Copper Independent (Independent) (32)

Loc: NE¼ sec. 19, (30-10E), at mouth of Silver Gulch, ½ mi. SE. of Silverton. **Elev:** 3,000 ft. **Access:** Road from Silverton. **Prop:** 3 claims. **Owner:** Leased from Lambda Chemical Corp. by Ore Recoveries Corp. (1943). Copper Independent Consolidated Mining Co. (1902-1908). **Ore:** Gold, silver. **Ore min:** Pyrite, arsenopyrite. **Gangue:** Sheared granite and quartz. **Deposit:** Mineralized shear zone in granite contains ore bodies from 100 to 200 ft. in dia. and 2 to 3 ft. wide. **Dev:** 3 adits, drifts, and raises totaling about 700 ft. **Assays:** 0.6 to 0.7 oz. Au. **Prod:** 1 carload of picked ore shipped to Everett smelter prior to 1901. **Ref: 14**, pp. 39-40. **22**, p. 9. **33**, 1907, p. 518; 1908, p. 557. **43**, 1902, p. 832. **46**, pp. 164-165. **63**, p. 18. **88**, p. 68. **114**, no. 5, 1909, pp. 98-99. **158**.

Dahl

(See Eclipse.)

Del Campo

Loc: On the E. slope of the ridge dividing the Sauk, Sultan, and Stilaguamish watersheds, overlooking Crater Lk., 2½ mi. W. of Monte Cristo. **Prop:** 3 claims. **Owner:** Del Campo Metals Co. **Ore:** Gold, silver, copper. **Ore min:** Chalcopyrite. **Deposit:** 10- to 30-ft. fracture zone that can be traced for 2,000 ft. on the surface. **Dev:** 25-ft. adit, several open cuts. **Assays:** Surface ore runs $44.86 Au, Ag, 13.8% Cu. **Ref: 14**, p. 23. **63**, p. 14. **105**, 1909, p. 871.

Eclipse (Dahl) (33)

Loc: Secs. 17, 18, and 19, (30-10E), on S. side of Stilaguamish R. at foot of Huckleberry Ridge, Silverton dist. **Prop:** 51 claims (1952). **Owner:** R. D. Taft, Robe, Wash. (1952 ——). Eclipse Consolidated Mining & Investment Co. of Seattle (1915-1918). Virginia-Agenda Co. (1921-1924). Chemical Products Ass'n. (1925-1926). Lambda Chemical Corp. leasing to Ore Recoveries Corp. (1943). **Ore:** Gold, silver, copper, mercury. **Ore min:** Arsenopyrite, native mercury. **Deposit:** 1 vein carrying arsenopyrite. 1 vein in Marble Gulch carrying gold, silver, and copper. Mercury in some workings. **Dev:** More than 5 mi. of underground work reported. **Assays:** Reported to be 7% Hg from an incline and 3% Hg from an adit just S. of the incline. **Ref: 14,** pp. 42-43. **63,** p. 19. **97,** 1913, p. 210. **98,** 1918-1926. **105,** 1915, p. 742. **114,** no. 5, 1909. **141,** p. 88. **158.**

Edison (49)

(See also Mineral Center.)

Loc: NE¼ sec. 29, 29-11E), Silver Cr. dist. **Owner:** Mineral Center Mining Co., Tacoma, Wash. (1951). Bonanza Mining & Smelting Co. (1901). **Ore:** Gold, copper, silver. **Ore min:** Chalcopyrite, arsenopyrite. **Deposit:** 50-ft. vein in granite. **Assays:** $10 to $15 per ton. **Ref: 63,** p. 28. **88,** p. 74. **97,** 1907, p. 475. **98,** 1918, p. 60. **114,** no. 5, 1909.

"45" (Magus) (34)

Loc: Secs. 29 and 30, (30-10E). **Elev:** 2,500 to 4,500 ft. **Access:** 6 mi. from Sultan Basin road by road up Williamson Cr. **Prop:** 25 patented claims and fractions. **Owner:** The 45 Mines Inc. (1931-1943). Forty-five Consolidated Mining Co. (1902-1907). **Ore:** Silver, lead, gold, zinc, copper. **Ore min:** Galena, sphalerite, ruby silver, chalcopyrite, arsenopyrite, pyrite, pyrrhotite, marcasite, scheelite, tetrahedrite. **Gangue:** Quartz, calcite. **Deposit:** Mineralized fracture zones in metamorphic rocks. One vein, the Magus, has an indicated length of 3,000 ft. **Dev:** More than 4,000 ft. of underground workings. **Improv:** 3 cabins, the remains of several other buildings, and 2 tram lines. **Assays:** Ore shipped returned 0.35 to 1.06 oz. Au, 48.4 to 171.4 oz. Ag. Other assays show 3.7% to 43% Zn, 4.6% to 6.5% Pb, 6.2% to 18.5% As, 0.28 to 0.6 oz. Au, 8.0 to 10.4 oz. Ag. **Prod:** Approx. 3,185 tons of ore 1896 to 1902. **Ref: 14,** p. 47. **23,** pp. 39-46. **33,** 1907, p. 605. **43,** no. 5, 1898, pp. 39-40. **46,** pp. 166-167. **63,** p. 24. **88,** pp. 66-67. **104,** 9/15/35, p. 23. **105,** no. 2, 1905, p. 35. **106,** 5/21/31; 9/3/31, p. 13. **112,** p. 179. **114,** no. 5, 1909. **129,** pp. 296-297. **158. 159,** p. 136.

Glory of the Mountain (35)

Loc: SE¼ sec. 2, (29-11E), ½ mi. NE. of Goat Lk., Monte Cristo dist. **Prop:** 7 claims. **Ore:** Gold, silver. **Ore min:** Arsenopyrite,

pyrite. **Deposit:** Wide mineralized fracture zone contains an ore body 20 ft. wide. **Dev:** 40-ft. adit. **Assays:** $21 to $27 Au, $1.80 to $4.80 Ag. **Prod:** Has produced. **Ref: 14,** p. 23. **58,** p. 24. **63,** p. 16.

Golden Chord

(See Justice.)

Good Hope (39)

Loc: Sec. 33, (29-11E), 2 mi. from Monte Cristo. **Access:** Road from Mineral City. **Prop:** 10 patented claims: 41144 and Fraction, Evening Star, Lucky Boy, Good Hope, Starlight, Starlight Ext. No. 2, Prince, Good Hope Ext., Queen. **Owner:** F. P. Hallinan, Portland, Oreg. (1941). Good Hope Gold & Copper Mining Co. (1909). **Ore:** Gold, silver, copper, lead, zinc, vanadium. **Dev:** About 2,000 ft. of adits. **Prod:** Reportedly 1 carload in 1909. **Ref: 58,** p. 26. **114,** no. 5, 1909. **158.**

Great Northern (47)

Loc: NW¼ sec. 35, (29-8E), on S. side of Sultan R. **Access:** Road up Sultan R. **Prop:** 3 claims. **Ore:** Gold, copper, silver. **Ore min:** Chalcopyrite, pyrite. **Deposit:** Mineralized shear zone 60 ft. wide along a contact between granite and slate. Ore minerals are disseminated throughout the zone. **Assays:** Av. of 6 assays showed $32 Au, $3.45 Cu, $1.76 Ag. **Ref: 14,** pp. 47-48. **63,** p. 25.

Independent

(See Copper Independent.)

Index Gold Mines, Inc. (45)

Loc: Secs. 18 and 19, (28-11E), Silver Cr. dist. **Access:** Road. **Prop:** 5 unpatented claims: Daily West, Daily West No. 1, LeRoy, Snow Shoe, Virtue. **Owner:** Henry Siepmann, Seattle, Wash. (1942). **Ore:** Gold, silver, lead, zinc. **Ore min:** Arsenopyrite. **Deposit:** Said to be a vein more than 18 in. wide. **Dev:** 300-ft. adit. **Assays:** Typical assay said to be 5.4 oz. Ag, 6.4% Pb, 2.5% Zn, 11.59% As. **Prod:** Reportedly shipped 10 tons of ore per day in 1939. **Ref: 14,** p. 28. **46,** p. 168. **158.**

Iowa

(See Washington-Iowa.)

Jasperson (Clara Thompson, Webster, Commonwealth, Mc-Combs) (44)

Loc: Sec. 36, (29-10E) and sec. 6, (28-11E), N. of the Bear claim on S. slope of Mineral Mtn., on W. side of the W. Fk. of Silver Cr. **Elev:** 3,400 ft. **Access:** Truck road. **Prop:** 17 unpatented claims, including: Sigma, Bullion King. **Owner:** Leased by W. Eggart, Seattle, Wash. (1941). Mrs. Clara Thompson and

Lawrence Thompson, Index, Wash. (1938). McCombs & Co. (1934). **Ore:** Gold, silver, copper, lead, mercury, vanadium. **Ore min:** Cinnabar, pyrite, arsenopyrite, chalcopyrite, galena, sphalerite, stibnite, vanadinite. **Gangue:** Quartz. **Deposit:** Narrow mineralized fracture zones in granodiorite. Ore shoots are sporadic and only 3 to 8 in. wide. **Dev:** Adit more than 30 ft. long and other workings said to total 3,000 ft. **Assays:** 8 samples show 0.02 to 0.44 oz. Au, tr. to 19.20 oz. Ag, tr. to 4.5% Cu, 0.8% to 15.1% Pb. **Ref:** **13**, p. 160. **14**, pp. 30, 34, 38. **58**, p. 72. **63**, p. 30. **158**.

Justice (Golden Chord) (40)

Loc: Sec. 27, (29-11E), on NW. side of Wilmon Peak overlooking Monte Cristo. **Elev:** 4,000 ft. **Access:** ¾ mi. by trail from Monte Cristo. **Prop:** 3 claims: John, Thomas, Irma. **Owner:** John Birney, Everett, Wash. (1940). Justice Mining Co. (1905). **Ore:** Gold, silver. **Ore min:** Arsenopyrite, pyrite. **Gangue:** Quartz. **Deposit:** A 1- to 3-ft. mineralized shear zone in andesite contains ore bodies from 70 to 400 ft. in length. Nearly all the known ore has been removed. **Dev:** 3 adit levels totaling about 2,200 ft. **Assays:** Av. assays showed 0.43 oz. Au, 3 oz. Ag, 12.4% As. **Prod:** Considerable ore produced from 1903 to 1905. **Ref:** **14**, p. 20. **58**, p. 25. **63**, p. 13. **91**, p. 241. **97**, 1905, p. 337. **149**, pp. 813, 819. **158**.

Last Chance (46)

Loc: Sec. 18, (28-11E), on W. side of Silver Cr., 2 mi. N. of Galena. **Access:** Truck road. **Owner:** Silver Creek Mining Co. (1907-1935). **Ore:** Gold, silver, lead. **Deposit:** Nearly vertical vein that strikes NW. **Dev:** 1 crosscut adit. **Prod:** Small amount in 1935. **Ref:** **14**, p. 28. **33**, 1907, p. 1026. **63**, p. 29. **97**, 1935, p. 354. **158**.

Louise

(See Mineral Center.)

McCombs

(See Jasperson.)

Magus

(See "45".)

Mineral Center (Bonanza, Edison, Louise, Washington-Iowa) (50)

(See also Edison, Washington-Iowa.)

Loc: W½ sec. 32, (29-11E), ¼ mi. E. of Red Gulch on N. side of Silver Cr. **Elev:** 3,500 ft. **Access:** 1½ mi. by trail from Mineral City. **Prop:** 13 claims, 10 of which are patented: Rattler, Emma, White Rose, Jessie, Monarch, Red Rose, Louise, Louise Fr., Jumbo, Blackbeard, Blue Bird, Torino, Edison. **Owner:**

Mineral Center Mining Co., Tacoma, Wash. (1919-1951). Bonanza Mining & Smelting Co. (1902-1918). **Ore:** Gold, copper, silver, lead, zinc. **Ore min:** Pyrite, chalcopyrite, malachite, azurite, galena, arsenopyrite, pyrrhotite, sphalerite, molybdenite. **Gangue:** Silicified rock. **Deposit:** Mineralized shear zones in metasediments. One zone is 40 ft. wide and contains abundant pyrite and some chalcopyrite. **Dev:** ·Workings on 3 levels total 3,500 ft. **Improv:** 5 buildings, sawmill, compressor, blacksmith shop, and hand tools (1946). **Assays:** Various assays show from 0.04 to 0.18 oz. Au, 0.04 to 7.38 oz. Ag. The large zone assays 0.10 oz. Au, 7.38 oz. Ag. **Ref: 88**, p. 74. **97**, 1907, p. 475. **98**, 1918, p. 60. **150**, p. 36. **158.**

Monte Cristo (Mystery, Pride) (37)

Loc: Sec. 26, (29-11E), 1 mi. E. of Monte Cristo. **Elev:** 3,000 to 5,500 ft. **Access:** Trail from Monte Cristo. **Prop:** 14 claims, including: Pride of the Woods, Pride of the Mountains, Mystery, Galore, Eighty Nine, Baltic. **Owner:** H. S. Ofstie, Everett, Wash. (1949). Monte Cristo Mining & Concentration Co. (1902). Monte Cristo Metals Co. (1907). Puget Sound Reduction Co. (1907-1918). Boston-American Mining Co. (1922-1926). **Ore:** Gold, silver, lead, zinc, copper. **Ore min:** Arsenopyrite, pyrite, chalcopyrite, galena, sphalerite, jamesonite, realgar. **Gangue:** Quartz, calcite, sheared rock. **Deposit:** Shear zone in schist and tonolite contains lenses of sulfide ore from 100 to 300 ft. in dia. and 1 to 15 ft. thick. Mineralization in other minor fractures. **Dev:** 3 main levels of adits and several smaller adits. Total underground work about 12,000 ft. **Assays:** 0.20 oz. Au, 3.00 oz. Ag, more than 6% As. Other assays showed, for surface ores: 0.95 oz. Au, 12 oz. Ag, 4% Cu, 5% Pb; for deeper ores: 0.6 oz. Au, 7 oz. Ag, 1% Cu, 2½% Pb. **Prod:** 300,000 tons of ore. The ore came mainly from the Mystery and Pride claims above the 700-ft. level. **Ref: 13,** p. 148. **14**, p. 21. **33**, 1907, p. 941; 1908, p. 1142. **63**, pp. 12-13. **88**, pp. 71-72. **97**, 1907, p. 475. **98**, 1918-1926. **100**, 1897, p. 180. **112**, p. 197. **132**, pp. 130-134. **149**, pp. 803-804, 813, 818-821, 841. **158**. **159**, p. 136.

Mystery
(See Monte Cristo.)

North Star
(See Sunrise.)

Oldfield
(See Sunrise.)

Peabody (41)
(See also Sidney.)
Loc: NE¼ sec. 27, (29-11E), on W. slope of Wilmon Peak, Monte Cristo dist. **Elev:** 3,400 ft. **Access:** Trail from Monte

Cristo. **Prop:** 1 patented claim: Remnant; 6 unpatented claims: Gene Elton, Lincoln, Sheridan, Kittie Mackay, Grant, Sidney. **Owner:** Louis H. McRae, Everett, Wash. (1949). Kitty A. Peabody, Edmonds, Wash. (1936). **Ore:** Gold, silver, lead, zinc. **Ore min:** Arsenopyrite, pyrite, galena, sphalerite. **Gangue:** Quartz and altered andesite. **Deposit:** Mineralized fracture zone in andesite ranges from a few in. to 18 in. wide. **Dev:** 200-ft. adit, several short adits and open cuts. **Assays:** 0.3 to 0.6 oz. Au, 1 to 36 oz. Ag. **Prod:** 1905. **Ref:** **14,** p. 21. **97,** 1905, p. 336. **114,** no. 5, 1909, p. 103. **157. 158.**

Pride

(See Monte Cristo.)

Queen Anne (29)

Loc: SE¼ sec. 27, (32-9E), on Jumbo Mtn., Darrington dist. **Owner:** Ole and Pete Nesta, Darrington, Wash. (1942). **Ore:** Gold, silver, copper, lead, zinc. **Ore min:** Chalcopyrite, galena, sphalerite, arsenopyrite, bornite. **Gangue:** Quartz. **Deposit:** Ore minerals associated with quartz in a 3-ft. fracture zone in granite. **Dev:** 1 adit. **Assays:** Tr. to $550 Au, tr. to $60 Ag, tr. to $35 Cu. **Ref:** **14,** p. 10.

Rainy (Ben Lomond) (36)

Loc: Sec. 22, (29-11E), ¼ mi. NE. of Monte Cristo. **Elev:** 3,000 ft. **Access:** Trail from Monte Cristo. **Prop:** 3 claims. **Owner:** J. F. Birney, Everett, Wash. (1930-1940). Ben Lomond Mining Co. (1914-1915). **Ore:** Gold, silver. **Ore min:** Arsenopyrite, pyrite. **Deposit:** Mineralized fracture zone in schist and andesite. Main ore body is 300 ft. in dia. and 8 in. to 5 ft. wide. **Dev:** 800-ft. adit, 200-ft. shaft. **Assays:** 862 tons shipped to smelter in 1913 to 1915 av. 0.638 oz. Au, 2.20 oz. Ag, 19.6% As. **Prod:** 20,000 tons of ore reported. **Ref:** **13,** p. 149. **14,** pp. 21-22. **63,** p. 13. **97,** 1914, p. 651; 1915, p. 572. **149,** p. 814. **158.**

Sidney (42)

(See also Peabody.)

Loc: Sec. 27, (29-11E), Monte Cristo dist. **Prop:** 1 unpatented claim (part of Peabody group). **Owner:** L. H. McRae, Everett, Wash. (1950). **Ore:** Gold, silver, copper. **Ore min:** Pyrite, chalcopyrite. **Gangue:** Quartz. **Dev:** Small amount. **Ref:** **13,** p. 150. **14,** p. 25. **114,** no. 5, 1909, p. 104. **149,** p. 833. **157.**

Sunrise (North' Star, Oldfield) (30)

Loc: NE¼NE¼ sec. 29, (32-9E). **Access:** 5 mi. by road to railroad near Darrington. **Prop:** 4 claims. **Owner:** W. E. Oldfield, Seattle, Wash. (1951-1952). **Ore:** Gold, silver, copper, lead, zinc, molybdenum. **Ore min:** Chalcopyrite, galena, molybdenite, arsenopyrite, sphalerite, pyrite. **Gangue:** Quartz. **Deposit:** Narrow fractures cutting granite are filled with quartz and ore minerals. **Dev:** 2 adits aggregating 155 ft. **Assays:** 0.08 to 0.55 oz. Au, 1.3 to 9.1 oz. Ag, 0.8% to 1.67% Cu, 0.1% Mo, 9% As. **Ref: 14,** p. 10. **133,** p. 40.

Undaunted (51)

Loc: Sec. 31, (29-11E), on Mineral Mtn., Silver Cr. dist. **Prop:** 4 claims. **Ore:** Gold, silver. **Ore min:** Pyrite. **Deposit:** 6 mineralized shear zones that range from 1½ to 6 ft. in width and carry from 6 to 36 in. of ore. **Dev:** 35-ft. adit and several open cuts. **Assays:** $12 to $70 per ton. **Ref: 14,** p. 38. **63,** p. 29.

Washington-Iowa (52)

(See also Mineral Center.)

Loc: Sec. 32, (29-11E), Silver Cr. dist. **Owner:** Mineral Center Mining Co. (1924-1951). Washington-Iowa Copper Mining Co. (1907-1918). **Ore:** Copper, lead, zinc, gold, silver. **Ref:** 33, 1908, p. 1414. **98,** 1918-1926. **105,** 1910, p. 63. **112,** p. 209. **116,** no. 1, 1907, p. 18.

Wayside (31)

Loc: SE¼ sec. 8, (30-7E), 1½ mi. E. of Granite Falls. **Elev:** 1,200 to 1,500 ft. **Access:** 1½ mi. above the Yankee Boy property by road. 10 mi. from railroad at Hartford. **Prop:** 15 patented claims. **Owner:** F. G. DeShane, Seattle, Wash. (1943-1953). Wayside Mining Co. (1905). American Copper Co. (1924). Riverside Minerals Co. (1928). Vanguard Metals, Inc., Everett, Wash. (1936-1939). **Ore:** Copper, gold, silver, lead, zinc, vanadium. **Ore min:** Chalcopyrite, pyrite, galena, sphalerite, bornite. **Gangue:** Cherty quartz. **Deposit:** Vein 6 to 18 in. wide cutting slate and siliceous limestone. **Dev:** Shaft and 7 levels. 6 under water. **Assays:** 0.01 to 0.25 oz. Au, 6 to 10 oz. Ag, 10% Cu. The reported occurrence of vanadium has not been verified. **Prod:** About $500,000 worth of high-grade ore shipped. **Ref: 14,** p. 13. **40,** p. 37. **97,** 1905, 1919, 1924, 1928. **104,** 12/30/36, p. 29. **158.**

Webster

(See Jasperson.)

STEVENS COUNTY

Acme

(See F. H. and C.)

Antelope (246)

Loc: Sec. 20, (39-38E), Orient dist. **Elev:** 3,650 ft. **Access:** Near road. **Prop:** 1 unpatented claim. **Owner:** Abandoned (1941). **Ore:** Gold, copper. **Ore min:** Pyrite, pyrrhotite, chalcopyrite, melanterite. **Gangue:** Quartz. **Deposit:** 2½- to 4-ft. vein in Jumbo volcanics. **Dev:** Inclined shaft about 100 ft. deep. **Assays:** Sample taken across the vein said to assay $26.25. **Prod:** 50 tons of ore said to have been mined. **Ref: 30,** p. 135. **164,** p. 251.

Beecher (235)

   Loc: Near center sec. 31, (40-37E), Orient dist. **Elev:** 2,500 ft. **Access:** 2 mi. NE. of Rockcut by road. **Owner:** Beecher Gold Mining Co. (1909-1924). **Ore:** Gold, silver, copper, lead. **Ore min:** Free gold, pyrite, sylvanite, galena, limonite, chalcopyrite. **Gangue:** Quartz, calcite. **Deposit:** 2 quartz veins, one 4 to 24 in. wide, another as much as 8 ft. wide. Country rock is schist and Rossland volcanics cut by diabase. **Dev:** 65-ft. shaft with 115-ft. drift at bottom. Other short drifts. **Assays:** Av. $4 to $24 Au, high-grade $136 to $296 Au. **Prod:** 2 shipments aggregating 22 tons prior to 1913. **Ref: 7,** p. 88. **97,** 1915, p. 575; 1916, p. 616. **98,** 1918, 1922, 1925. **114,** 8/09, p. 60. **164,** p. 285.

Benvenue

(See Gold Reef.)

Charity

(See F. H. and C.)

Easter Sunday (237)

   Loc: Near center E½ sec. 22, (40-37E), Orient dist. **Elev:** 3,550 ft. **Access:** 8 mi. E. of Rockcut on Easter Sunday Lk. road. **Prop:** 2 patented claims: Easter Sunday, Cairn. **Owner:** D. C. Ames & Associates, Kenilworth, Ill. (1949). Memphis Mining Co. (1903). Waukegan Mining Co. (1903). Waukegan & Washington Mining & Smelting Co. (1907). Forest Mining & Milling Co. (1909). **Ore:** Gold, copper, silver. **Ore min:** Pyrite, chalcopyrite, sphalerite, galena, boulangerite, tetrahedrite. **Deposit:** 2- to 6-ft. quartz vein in cherty argillite. Vein sparsely mineralized, and cut off by a monzonite dike. **Dev:** 130-ft. inclined shaft and 2 short levels run at 70 and 100 ft. from the collar total 550 ft. of workings. 480-ft. diamond drill hole. **Assays:** 20-ton shipment av. 1.21% Cu, 0.44 oz. Au, 12.4 oz. Ag. **Prod:** Shipped approx. 20 tons of ore in 1909. **Ref: 7,** pp. 83-84. **30,** p. 131. **33,** 1907, p. 1156. **98,** 1925, p. 1817. **100,** 1903, pp. 100-101. **111,** p. 8. **112,** p. 178. **114,** no. 5, 1909, p. 60. **116,** no. 12, 1907, p. 15. **132,** pp. 141-142. **164,** pp. 274-275.

Eureka (Eureka and Orient, Indian, Orient Eureka) (236)

   Loc: Secs. 13, 24, and 25, (40-36E), E. of Kettle R., Orient dist. **Elev:** 2,485 ft. **Access:** 5 mi. by road from railroad at Rockcut. **Prop:** 11 patented and 14 unpatented claims. **Owner:** Orient Eureka Gold Mines Co., Tacoma, Wash. (1951). **Ore:** Gold, silver, lead, zinc, copper. **Ore min:** Galena, chalcopyrite, sphalerite, pyrite. **Deposit:** Several quartz veins in schist and diabase. **Dev:** 1,500 ft. workings in adits, shafts, and open cuts. **Assays:** Some veins returned $32 Au, others $60 Pb and Zn. **Prod:** Produced in 1890's. **Ref:** 108, 9/39, p. 30. 164, p. 292.

Eureka and Orient
(See Eureka.)

F. H. and C. (Faith, Hope, and Charity; or Acme) (233)

Loc: Secs. 19 and 30, (40-37E), Orient dist. **Elev:** 2,600 ft. **Access:** 5 mi. NE. of Rockcut by road. **Prop:** 3 patented, 10 unpatented claims. **Owner:** Goldstake Mining Corp. (1941). F. H. & C. Gold Mining Co. (1918-1920). Acme Consolidated Mines, Inc. (1930). **Ore:** Gold, silver. **Ore min:** Pyrite. **Deposit:** 3- to 5-ft. quartz vein. **Dev:** 250-ft. shaft with drifts on 4 levels, total 1,500 ft. Large amount of open-cut work. **Assays:** Smelter shipment said to return 2.36 oz. Au, 1.9 oz. Ag. **Prod:** 1 small shipment to Tacoma smelter about 1917, produced 1929. **Ref:** 1-B, 7/21/32, p. 1. 30, p. 118. 97, 1918, p. 510; 1930, p. 667. 98, 1920-1925. 104, 6/15/35, p. 27; 2/29/36, p. 28; 12/15/36, p. 27. 112, p. 178. 113, April-May 1934, p. 15. 158. 164, pp. 277-278.

Faith
(See F. H. and C.)

First Thought (239)

Loc: Secs. 7 and 18, (39-37E). **Elev:** 2,900 ft. **Access:** 3 mi. E. of railroad at Orient, on First Thought Mtn. Lookout road. **Prop:** 17 patented, 11 unpatented claims. **Owner:** Gold Syndicate Corp., Spokane, Wash. (1949 ——). First Thought Gold Mines Co., Ltd. (1905-1934). First Thought Mining Corp. (1937-1944). **Ore:** Gold, silver. **Ore min:** Pyrite, free gold. **Gangue:** Quartz, calcite. **Deposit:** Mineralized zone 15 to 110 ft. wide in rhyolite porphyry and quartz latite. Richer ores in the zone occur at intersections of faults. **Dev:** 1½ mi. of workings on 7 levels. **Assays:** 40,000 tons of ore av. ¾ oz. Au, ½ oz. Ag. **Prod:** $1,350,000 in gold prior to 1948. Produced 1904, 1906-1909, 1934-1942. **Ref:** 7, pp. 71-76. 30, pp. 117-118. 39. 68, p. 9. 97, 1905, 1907-1910, 1931, 1935, 1937-1941. 98, 1918-1926. 99, 11/27/34. 104, 6/15/32, p. 27; 7/15/34, p. 26. 106, 5/5/32. 112, p. 178. 113, 2/18/37, p. 8. 129, p. 164. 130, p. 67. 141, p. 22. 158. 164, pp. 258-260.

First Thought Extension (240)

Loc: Sec. 18, (39-37E), SE. of the First Thought property, Orient dist. **Access:** Near road. **Prop:** Several claims. **Owner:** May be part of holdings of the Gold Syndicate Corp. **Ore:** Gold. **Ore min:** Pyrite. **Deposit:** Latite flows cut by monzonite dikes. Numerous fault zones in both formations are mineralized by pyrite said to carry gold. **Dev:** 650 ft. of work in main adit. **Assays:** Said to assay up to $400 Au. **Ref: 30,** p. 135. **114,** no. 5, 1909, p. 59. **164,** p. 261.

Gem (244)

Loc: Sec. 19, (39-37E), on N. slope of First Thought Mtn., 6 mi. NE. of Orient. **Elev:** 3,440 ft. **Access:** Road. **Prop:** 4 unpatented claims. **Owner:** Paul Anderson, Colville, Wash. (1941). **Ore:** Gold. **Ore min:** Auriferous pyrite. **Gangue:** Quartz. **Deposit:** Lavas and interbedded sediments cut by monzonite and fine-grained granite dikes. Ore occurs in a 4-ft. fracture zone in one of the granite dikes. **Dev:** 250-ft. adit, 105-ft. shaft, and 4 drifts from 12 to 60 ft. long. **Assays:** Some selected samples assayed $1,000 Au. Ore said to av. $40 per ton. **Prod:** 1939. **Ref: 30,** p. 123. **97,** 1940, p. 480. **106,** 7/21/32, p. 1. **158. 164,** p. 271.

Gold Bar (248)

Loc: Secs. 15 and 22, (37-38E). **Elev:** 1,600 to 2,100 ft. **Access:** 1,000 ft. E. of State Highway 22 by road. 1½ mi. to railroad at Evans. **Prop:** 120 acres of deeded land in 8 patented claims: Mayflower Nos. 1 to 6 and Contact Nos. 1 and 2. **Owner:** Gold Bar Mining Co., Bossburg, Wash. (1948-1950). **Ore:** Gold, silver, copper, lead, zinc. **Ore min:** Pyrite, tetrahedrite, sphalerite. **Deposit:** Quartz veins from ½ in. to 8 in. wide filling fractures in quartzite and argillite. Veins sparsely mineralized. Small tonnage of ore exposed. **Dev:** About 1,000 ft. of workings in 4 adits and 2 inclines. **Assays:** 7 assays show from 0.005 to 0.15 oz. Au, 1.2 to 62.0 oz. Ag, tr. to 1.3% Pb, tr. to 1.7% Zn, tr. to 0.3% Cu, 1.02% to 2.70% As. **Prod:** 400-lb. test shipment to Bunker Hill smelter prior to 1945. **Ref: 30,** pp. 80-81. **69,** p. 8. **158.**

Gold Ledge (249)

Loc: SW¼ sec. 4, (36-38E), Kettle Falls dist. **Access:** Near road. **Prop:** 460 acres of deeded land. **Owner:** Franklin Good (1941). Silver Queen Mining Co. (1934-1935). **Ore:** Gold, silver. **Deposit:** Quartz vein along contact of argillite and porphyry. **Dev:** 150-ft. inclined shaft, large open cut. **Assays:** 6 tons shipped netted $40 per ton Au, Ag. **Prod:** 6 tons in 1934. **Ref: 30,** p. 56. **97,** 1935, p. 355. **99,** 1/8/35. **104,** 12/15/34, p. 22. **113,** April-May 1934, p. 5.

Gold Reef (Benvenue, Golden Reef) (250)

Loc: SW¼SE¼ sec. 9, (36-38E), on top of hill, Kettle Falls dist. **Elev:** 3,100 ft. **Access:** Road. **Prop:** 4 unpatented claims. **Owner:** William Setzen and Ross Morehead, Colville, Wash. (1949). Bunker Hill & Sullivan Mining & Concentrating Co. (1932). M. I. Stellman and G. F. Grundy (1938). **Ore:** Gold, silver, copper. **Deposit:** Quartz vein said to av. 2 to 3 ft. wide and to be traceable for 800 ft. Vein is along contact between argillite and acidic dike. **Dev:** 750 ft. of adits and 165 ft. of shafts and inclines. **Improv:** Small homemade mill. **Assays:** 24 tons shipped to smelter assayed 0.542 oz. Au, 0.834 oz. Ag. **Prod:** $100,000 reported prior to 1935. 1939 (24 tons). 2 or 3 carloads in 1946 and 1947. **Ref: 30,** p. 55. **68,** p. 9. **97,** 1910-1912, 1915. **104,** 3/15/32, p. 27; 8/30/32, p. 28. **106,** 4/7/32, p. 10. **158. 164,** pp. 231-232.

Golden Hope
(See Sunday.)

Golden Reef
(See Gold Reef.)

Homestake (245)

Loc: Sec. 19, (39-38E). **Elev:** 3,800 ft. **Access:** Near road. **Prop:** Several unpatented claims. **Owner:** Julius Weston, Northport, Wash. (1941). Minorca-Homestake Mines Co. (1922). Centennial Mines Co. (1924-1926). **Ore:** Gold, copper, silver, lead. **Ore min:** Chalcopyrite, pyrite, pyrrhotite. **Gangue:** Quartz. **Deposit:** 4½-ft. vein in Jumbo volcanics. **Dev:** 2 shafts 27 and 12 ft. deep. Trenching for 100 ft. along the vein. **Assays:** Said to assay $17.00 Au, $8.60 Cu, $3.15 Ag. At one place the vein is reported to show 4½ ft. of $31 ore. **Prod:** 100 tons of ore shipped. **Ref: 30,** p. 136. **98,** 1925, p. 1808; 1926, p. 1573. **164,** pp. 251-252.

Hope
(See F. H. and C.)

Indian
(See Eureka.)

Kettle River
(See White Elephant.)

Michigan (241)

Loc: Secs. 7 and 18, (39-37E), Orient dist. **Access:** 1 mi. from road. 4 mi. from Orient. **Prop:** 4 patented claims: Climax, Plutania, Moonlight, Butte. **Owner:** Abandoned (1941). Michigan Gold Mining Co. (1922). Treasure Gold Mining Co. (1937). **Ore:** Gold, silver. **Ore min:** Pyrite. **Deposit:** Gold-bearing pyrite occurs along zones of crushing and faulting in latite flows. The latite has been cut by numerous dikes of monzonite porphyry. **Dev:** 800-ft. and 400-ft. adits, shallow shaft. **Ref: 30.** p. 133. **98,** 1920, p. 1497; 1925, p. 1824. **114,** no. 5, 1909, p. 59. **158. 164,** pp. 260-261.

Orient Eureka
(See Eureka.)

St. Crispin (247)

Loc: Sec. 25, (40-39E), on Sheep Cr. **Elev:** 1,400 ft. **Access:** About 1 mi. W. of Northport by road. **Prop:** 6 unpatented claims. **Owner:** Abandoned (1941). St. Crispin Mining, Milling, Smelting and Development Co. (1908-1926). **Ore:** Gold, copper, silver, lead. **Gangue:** Quartz. **Deposit:** Mineralized zone in argillite. Said to be 8 more similar zones. **Dev:** 470-ft. shaft, 50-ft. shaft, and about 150 ft. of crosscuts, all caved (1941). **Assays:** 4% Cu, $26 Au, 13 oz. Ag. **Ref:** **30,** p. 96. **98,** 1925, p. 1830; 1926, p. 1593. **116,** no. 2, 1908, p. 24. **164,** pp. 315-316.

Second Thought (243)

Loc: Near center SW¼ sec. 18, (39-37E), adjoining First Thought property on SE. **Elev:** 2,700 ft. **Access:** 3 mi. E. of Orient by road. **Prop:** 5 claims and 3 fractions. May be part of holdings of Gold Syndicate Corp. **Owner:** Second Thought Gold Mines Co. (1909-1920). Valley Dew M. & M. Co. (1908). **Ore:** Gold, silver. **Ore min:** Pyrite. **Deposit:** Pyrite impregnations and joint and fault fissure fillings in quartz latite cut by quartz monzonite dikes. **Dev:** 64-ft. shaft, 40-ft. shaft, 22-ft. shaft, 5 other shafts, and some open cuts. **Assays:** Pyritiferous material said to assay $2 Au; some surface ore from Searchlight claim assayed $8 Au. **Ref:** **7,** pp. 76-77. **30,** p. 121. **58,** p. 61. **114,** no. 5, 1909, p. 60. **116,** no. 4, 1908, p. 91. **164,** pp. 263-264.

Sunday (Sunday Morning Star, Golden Hope) (251)

Loc: S½SE¼ sec. 7, (36-38E), 2 mi. NE. of Kettle Falls. **Elev:** 2,000 to 2,300 ft. **Access:** Road. **Prop:** 3 unpatented claims: John G. Mine, Columbia Mines, Rock Rose Mine. **Owner:** Golden Hope Mining Co. (1940). Reportedly abandoned in 1941. **Ore:** Gold, silver. **Deposit:** 2 quartz veins 4 to 6 in. wide in argillite and limestone. **Dev:** 3 adits, one 350 ft. long; shaft 150 ft. deep; other shallow shafts and open cuts. **Assays:** Ore said to av. $25 per ton. **Prod:** 1912, 1915. **Ref:** **30,** p. 55. **97,** 1912, p. 923; 1915, p. 574. **158. 164,** p. 233.

Sunday Morning Star
(See Sunday.)

Swamp King (234)

Loc: Sec. 30, (40-37E), adjoining Little Giant on the E. **Elev:** 2,650 ft. **Access:** Road. **Prop:** 5 unpatented claims. **Owner:** Andrew Abrahamson and Joe Bodbout, Orient, Wash. (1941). Swamp King Mining Co. (1915). **Ore:** Gold, silver. **Ore min:** Free gold, pyrite. **Deposit:** Quartz vein about 8 in. wide in diabase cut by camptonite dikes. **Dev:** 60-ft. adit and 3 shafts total 200 ft. of workings. **Prod:** Small shipment reported. **Ref:** **7,** p. 88. **30,** p. 122. **158. 164,** pp. 285-286.

Titanic (Valley Dew) (238)

Loc: Sec. 7, (39-37E), W. of First Thought Mtn., Orient dist. **Elev:** 3,500 ft. **Access:** Road. **Prop:** 10 unpatented claims. **Owner:** Abandoned (1941). Titanic Mining Co., Spokane, Wash. (1915-1920). **Ore:** Gold, silver. **Ore min:** Pyrite. **Deposit:** Latite cut by numerous monzonite dikes. Fracture zones in these rocks are mineralized by auriferous pyrite. **Dev:** 551-ft. adit, 540-ft. adit, and 400-ft. Gettern adit. **Assays:** One zone said to assay $8 Au. Assay in 1909 reports $18.85 Au, 1 oz. Ag. **Ref: 30,** pp. 133-134. **114,** no. 5, 1909, p. 59. **164,** pp. 262-263.

Trophy (242)

Loc: Sec. 18, (39-37E), Orient dist. **Access:** Near road. **Prop:** 4 claims: Buckeye, Wild Rose, Wild Rose Fraction, Imperial. **Owner:** L. E. Gourlie, Orient, Wash. (1941). Valley Dew M. & M. Co. (1908). Trophy Gold Mining Co. (1909). Trophy Gold Mining & Milling Co. (1920). **Ore:** Gold. **Ore min:** Pyrite. **Deposit:** Latite and interbedded shale cut by monzonite porphyry dikes. Fracture zones mineralized with pyrite carry some gold. **Dev:** 300 ft. of adit and 100 ft. of shaft work. **Assays:** Said to range from tr. to $12 Au. **Ref: 30,** p. 123. **116,** no. 4, 1908, p. 91; no. 5, 1909, p. 60. **164,** pp. 265-266.

Valley Dew

(See Titanic.)

White Elephant (Kettle River) (232)

Loc: Near center sec. 19, (40-37E), Orient dist. **Elev:** 2,650 ft. **Access:** 4 mi. from Rockcut by road. **Prop:** 6 unpatented claims: Kettle River, Rossland Mountain, Golden Cycle, Golden Triangle, Little Tiger, Little Penny. **Owner:** Kettle River Gold Mining Co., Spokane, Wash. (1934-1941). Orient Gold Mines, Ltd. (1915-1918). Orient Golden Rock Mining Co. (1918-1922). Gold Fissure Mining Co. (1935-1937). **Ore:** Gold, silver, copper. **Ore min:** Pyrite. **Deposit:** Mineralized zone in limestone and quartzite cut by fine-grained dikes and quartz veins. **Dev:** 225-ft. shaft; 85-ft. crosscut and 110 ft. of drifting on 100-ft. level; 50 ft. of drifting and 50 ft. of crosscutting on 200-ft. level. **Assays:** Ore said to av. $7.00 Au. **Ref: 30,** p. 120. **98,** 1918-1925. **99,** 1/29/35. **104,** 6/15/35, p. 27; 2/29/36, p. 28; 12/15/36, p. 27; 2/28/37, p. 28. **106,** 7/21/32, p. 1. **112,** p. 196. **150,** p. 34. **158. 164,** p. 292.

WHATCOM COUNTY

Allen Basin (15)

Loc: Sec. 34, (38-17E), in Allen Basin, Slate Cr. dist. **Access:** 5 mi. by road NW. of Harts Pass. **Prop:** 14 patented claims. **Owner:** Slate Creek Mining Co., Seattle, Wash. (1949). Cascadian Engineering Corp. (1932, 1934). Flying Cloud Mining Co. (1932-1933). **Ore:** Gold, silver, lead, zinc. **Ore min:** Free gold, arsenopyrite, pyrite, and some galena and sphalerite. **Deposit:** At least 2 quartz veins 3 to 30 in. thick cutting sedimentary and igneous rocks. **Dev:** 2 shafts, a caved adit, and several open cuts. **Assays:** Av. grade of ore is $1.25 to $3.00 Au. All values av. $12 to $15 per ton. **Prod:** Some ore apparently shipped in early 1900's and a few tons in 1938-1940. **Ref:** **104,** 10/15/32, p. 29; 3/15/33, p. 17. **106,** 7/7/32; 10/6/32, p. 9; 11/6/32, p. 9. **158.**

Anacortes (14)

Loc: SW¼ sec. 24, (38-16E), near head of Cascade Cr., Slate Cr. dist. **Prop:** 7 claims. **Owner:** Anacortes Gold Mining Co. (1935-1939). **Ore:** Gold, silver. **Ore min:** Free gold, tellurides, sulfides. **Deposit:** Quartz vein av. 2 ft. wide in slate and conglomerate. **Dev:** 310-ft. drift, 100-ft. drift, 90-ft. drift. **Assays:** Av. $18 to $20 Au, Ag. **Prod:** Has produced. **Ref:** **63,** p. 57. **88,** pp. 49-50. **104,** 7/15/32, p. 29; 10/15/35, pp. 25-26; 9/15/36, p. 34. **113,** 7/36. **114,** no. 5, 1909.

Anoka (22)

Loc: Sec. 11, (37-16E), Slate Cr. dist. **Prop:** 6 patented claims. **Owner:** John E. Wells, Tulsa, Oklahoma (1952). Mountain Top Mining Co. (1912). **Ore:** Gold. **Assays:** As high as $48 Au. **Prod:** Stamp mill operated in 1900's. **Ref:** **158.**

Azurite (25)

Loc: Secs. 30 and 31, (37-17E) and secs. 25 and 36, (37-16E), on Mill Cr., Slate Cr. dist. **Access:** Truck road over Harts Pass from Methow Valley. **Prop:** 42 unpatented claims including Azurite, Copper Dike, Seattle (1949). 11 unpatented claims (1952). **Owner:** Azurite Gold Co., Winthrop, Wash. (1925 ——). Leased by American Smelting & Refining Co. (1934-1943). **Ore:** Gold, copper, zinc, silver, lead. **Ore. min:** Chalcopyrite, sphalerite, pyrite, pyrrhotite, galena. **Deposit:** A 2- to 7-ft. quartz vein cutting through argillite is mineralized in places. Other undeveloped veins are also reported. Ore-producing body has been depleted. **Dev:** About 3,000 ft. of adit work, chiefly on 2 levels. Since 1941 a 125-ft. winze was sunk and crosscuts were driven both ways from the winze. **Assays:** 58,358 tons shipped av. 0.473 oz. Au. **Prod:** 1920, 1930, 1934, 3,000 tons in 1936, 27,530 tons in 1937, 36,515 tons in 1938, 5,375 tons in 1939. Produced 1941. Gross value of production reported $972,000. **Ref:** **46,** pp. 230-231. **97,** 1920, 1930, 1934, 1937-1939. **98,** 1920-1926. **104,** 11/30/35, p. 26; 2/29/36, p. 28; 12/15/36, p. 27. **106,** 12/19/29; 5/1/30; 12/4/30; 3/19/31, pp. 19-20. **108,** 7/39, p. 22; 10/39, p. 32. **158.**

Beck and Short Grub (16)

Loc: Sec. 36, (38-17E), Slate Cr. dist. **Prop:** 2 claims. **Owner:** Mrs. John A. Barron, Anacortes, Wash. (1952). **Ore:** Gold, silver, copper. **Deposit:** Quartz vein 3 to 6 in. wide in slate intruded by porphyry. **Assays:** 2.75 oz. Au, 51.0 oz. Ag. **Ref:** 58, p. 6. 63, p. 57. 114, no. 5, 1909. 158.

Bonita
(See New Light.)

Boundary Gold
(See Lone Jack.)

Boundary Red Mountain (Red Mountain) (7)

Loc: NE¼ sec. 4, (40-9E), 2 mi. S. of the international boundary, Mt. Baker dist. **Access:** Road and trail from Chilliwack, B. C. **Prop:** 6 patented and 5 unpatented claims. Those patented are: Klondike, Rocky Draw, Mountain Boy Lode, Glacier, Climax, Climax Ext. No. 1. **Owner:** Gold Basin Mining and Development Co., Bellingham, Wash. (1954 ——). Judge Elmon Scott (1897-1912). Boundary Red Mountain Mining Co. (1915-1934). International Gold Mines, Ltd. (1935-1938). J. W. Langley, Sumas, Wash. (1943). Tom Bourn, Bremerton, Wash. (1952). **Ore:** Gold. **Ore min:** Free gold, pyrite, chalcopyrite, pyrrhotite. **Deposit:** Finely divided gold in a quartz vein which varies from 6 in. to 7 ft. and av. 3 ft. in width. The vein can be traced for 4,500 ft. on the surface. Diorite country rock. **Dev:** 3,000 ft. of drifts equally divided among 3 levels. Ore developed to depth of 433 ft. below No. 1 level and 100 ft. above. **Assays:** Mine-run ore. av. $15 Au. An ore shoot 520 ft. long and av. 26 in. wide had an av. value of 1.13 oz. Au. **Prod:** 1912-1917, ($148,578 in 1916, $131,918 in 1917), 1920-1922, ($95,000 in 1922), 1925, 1929-1930, 1935, 1936 (200 tons), 1937-1942, 1947. **Ref:** 43, 3/31/23, p. 597. 97, 1913-1916, 1921-1923, 1925, 1930, 1931, 1937-1941. 98, 1918-1926. 99, 12/18/34, 1/8/35, 3/12/35. 104, 4/15/35, p. 23; 5/30/35; 10/30/35, p. 25. 106, no. 18, 1922, pp. 1-2. 117, no. 18, pp. 1-2. 129, pp. 306-309. 130, p. 69. 141, pp. 20, 22, 34. 158.

Brooks-Willis
(See Lone Jack.)

Chancellor (Indiana) (17)

Loc: SW¼ sec. 26, (38-17E), 2 mi. N. of the New Light mill, Slate Cr. dist. **Access:** About 6 mi. NW. of Harts Pass by road. **Prop:** 4 patented claims: Indiana, Illinois, Ptarmigan, Chancellor. **Owner:** Frank D. Hyde, Berlin, Maryland (1946). Chancellor Mining Co. (1907). **Ore:** Gold, silver, lead, zinc. **Ore min:** Pyrite, arsenopyrite, galena, sphalerite. **Deposit:** Quartz-calcite veinlets fill fractures along shear zones in quartzite. Zones are 3 to 12 in. wide. **Dev:** 2 levels 250 ft. apart; one is a 225-ft. adit and the other is an adit with 650 ft. of drifts and crosscuts. **Assays:** 119.4 tons shipped av. 0.735 oz. Au, 3.48 oz. Ag. **Prod:** 1903, 119.4 tons in 1935-1939. **Ref: 58,** p. 33. **104,** 8/30/36, p. 26; 10/30/36, p. 32. **114,** no. 2, 1906, p. 22; no. 5, 1909, p. 90. **158. 159,** p. 133.

Eureka
(See New Light.)

Evergreen (8)

Loc: NW¼ sec. 21, (40-9E), on Swamp Cr., 2 mi. N. of Shuksan Cabins. **Elev:** 200 ft. above Swamp Cr. **Access:** Across the creek from Swamp Cr. road. **Prop:** 8 unpatented claims: Evergreen Nos. 6 to 13. **Owner:** Bert Lowry, Bellingham, Wash. (1943). Evergreen Mines, Inc., Seattle, Wash. (1937). **Ore:** Gold, silver, lead, zinc, copper. **Ore min:** Sphalerite, galena, pyrite, chalcopyrite. **Gangue:** Quartz, calcite. **Deposit:** Small ore stringers from ⅛ in. to 8 in. wide in argillite. Pyrite is also disseminated in the argillite. **Dev:** 2 drifts 40 and 30 ft. long. **Assays:** Ore said to av. $17.00 in all values. **Prod:** Shipped to Tacoma smelter in 1938. **Ref: 158.**

Galena
(See Verona.)

Gargett (5)

Loc: SW¼ sec. 4 and SE¼ sec. 8, (40-9E), on S. side of Red Mtn., Mt. Baker dist. **Elev:** 5,000 and 7,000 ft. **Access:** Trail from Twin Lakes. **Prop:** 7 unpatented claims. **Owner:** John Curtis, Victor Todd, Bert Boyer, Bellingham, Wash. (1952). Gargett Bros., Sumas, Wash. (1940). **Ore:** Gold, silver, copper, lead, zinc. **Ore min:** Sphalerite, galena, chalcopyrite, pyrite, malachite, chalcocite, azurite, massicot, pyrrhotite. **Deposit:** Mineralized quartz vein in limestone and a band of mineralized siliceous limestone in argillite and quartzite. The vein av. 12 in. in width for a length of 20 ft. **Dev:** Open cut on upper property and a 2000-ft. adit on the lower property, also a second adit. **Assays:** A 5-ton test shipment av. 1.03 oz. Au, 4.34 oz. Ag, 0.78% Cu. 8% Pb and Zn reported from upper adit. **Prod:** A 5-ton test shipment was made to the Tacoma smelter in 1940. **Ref: 58,** p. 23. **158.**

Goat (18)
(See also New Light.)

Loc: SW¼ sec. 33, (38-17E), SW. of and adjacent to the Allen Basin claims, Slate Cr. dist. **Prop:** Part of the Slate Creek group (1950). **Owner:** Slate Creek Mining Co. (1950). **Ore:** Gold. **Prod:** Produced many years ago. **Ref: 158.**

Goat Mountain (11)

Loc: On Goat Mtn., 4 mi. from Shuksan and 1 mi. E. of Swamp Cr., Mt. Baker dist. Probably in sec. 28, (40-9E). **Prop:** 2 claims. **Ore:** Gold. **Ore min:** Pyrite, free gold. **Deposit:** Quartz vein 6 to 12 in. wide, 450 ft. long. **Dev:** Open cuts. **Assays:** Possibly $60 to $70 Au. **Prod:** Small amounts in 1902 and 1903. **Ref: 158.**

Gold Run (6)

Loc: SE¼ sec. 5, (40-9E), Mt. Baker dist. **Ore:** Gold, silver, copper. **Prod:** 1938. **Ref: 58,** p. 26. **97,** 1939, p. 492. **99,** 2/12/35. **104,** 1/30/35, p. 23.

Golden Arrow
(See Tacoma.)

Great Excelsior (Lincoln, President) (4)

Loc: Sec. 6, (39-8E), W. of Wells Cr. and S. of N. Fk. of Nooksack R. **Elev:** 1,400 to 2,500 ft. **Access:** 2 mi. by trail from the Nooksack Forest Camp. **Prop:** 10 claims and 2 fractions. **Owner:** Great Excelsior Mining Co., H. E. Barnes, Bellingham, Wash. (1902-1952). Hammond Mining Co. (1914-1926). **Ore:** Gold, silver. **Ore min:** Pyrite, chalcopyrite, arsenopyrite, galena, sphalerite, tellurides, native silver. **Gangue:** Quartz, dolomite. **Deposit:** Brecciated greenstone cemented by sulfides. 2 mineralized zones; one is 75 ft. wide and said to be all paying ore. **Dev:** Several adits and a large stope total several hundred ft. of workings. **Assays:** 64 samples showed tr. to 0.08 oz. Au, the av. being 0.02 to .03 oz. In some samples Ag av. 2.0 oz. **Prod:** 10.000 tons with net return of $20,276 to 1915. 305.095 tons of conc. shipped av. 1.857 oz. Au per ton. **Ref: 1,** 1914, pp. 51-53. **88,** pp. 43-44. **98,** 1918-1926. **112,** p. 182. **114,** no. 5, 1909, p. 87. **119,** no. 2, 1913, pp. 1-3. **145,** p. 96. **158. 159,** p. 132.

Indiana
(See Chancellor.)

Lambert
(See Lone Jack.)

Lincoln
(See Great Excelsior.)

Lone Jack (Mount Baker, Post-Lambert, Brooks-Willis, Boundary Gold) (9)

Loc: SE¼ sec. 15 and sec. 22, (40-9E), about 1 mi. SE. of Twin Lakes, near head of W. Fk. Silesia Cr. **Elev:** 4,000 to 6,000 ft. **Access:** Truck road from end of Twin Lks. road. **Prop:** 5 patented, 17 unpatented claims. **Owner:** R. J. Cole, Seattle, Wash., lessee (1951). Mt. Baker Mining Co. (1898-1915). Boundary Gold Co. (1915-1924). Brooks-Willis Metals, Inc. (1928).

Estate of P. R. Brooks (1940). **Ore:** Gold, silver. **Ore min:** Free gold, gold telluride. **Deposit:** Quartz vein in schist has an av. width of 30 in. and is traceable for 2,500 ft. Values are localized in payshoots. Gold mostly too finely divided to be seen with the naked eye, but many fancy specimens of free gold were taken out. **Assays:** Av. tenor of the ore said to be $15 to $35 Au. Picked samples ran as high as $850 per ton. **Prod:** About $275,000 prior to 1915. Produced 1915-1918. Total production reported to be 9,463 oz. Au, 1,961 oz. Ag. **Ref:** **43**, 1902, p. 379. **88**, pp. 42-43. **97**, 1915-1917, 1924, 1928, 1930. **98**, 1920-1925. **105**, 1903, p. 140. **112**, p. 171. **119**, 1913, pp. 1-2. **129**, p. 309. **130**, p. 69. **141**, p. 22. **145**, pp. 95-96. **150**, p. 35. **158**. **159**, pp. 131-132.

Mammoth (19)

Loc: Sec. 33, (38-17E), Slate Cr. dist. **Elev:** 5,500 to 6,500 ft. **Access:** About 5 mi. NW. of Harts Pass by road. **Prop:** 6 patented and several unpatented claims. **Owner:** Harts Pass Mining Co., Seattle, Wash. (1946). Owens Gold Mines (1935). Mammoth Gold Mines Co. (1935-1936). Mammoth Gold Mining Co. (1936). **Ore:** Gold, silver, lead, zinc. **Ore min:** Pyrite, arsenopyrite, galena, sphalerite, free gold, tellurides. **Deposit:** A 1- to 3-ft. quartz vein in argillite and quartzite carries sulfides together with gold and silver. **Dev:** 3 adits total about 2,250 ft. **Assays:** The 15,000 tons of ore produced av. $26.50 per ton. **Prod:** Reportedly produced 15,000 tons of ore prior to 1900 and 30,000 tons by 1942. Reportedly more than $1,000,000 total prior to 1942. **Ref:** **63**, p. 57. **88**, pp. 47-48. **97**, 1923, p. 416; 1935, p. 355. **99**, 2/12/35. **104**, 4/30/35, p. 29; 11/30/35, p. 26; 3/15/36, p. 23; 8/30/36, p. 24. **158**. **159**, p. 133.

Monica
(See New Light.)

Mount Baker
(See Lone Jack.)

New Light (Eureka, Bonita, Slate Creek, Monica) (20)
(See also Goat.)

Loc: Center S½ sec. 27, (38-17E), about 7 mi. NW. of Harts Pass, Slate Cr. dist. **Elev:** 5,500 to 6,600 ft. **Access:** Road to mine from Mazama. **Prop:** 16 unpatented claims. **Owner:** Western Gold Mining Inc., Seattle, Wash. (1952——). Eureka Mining Co. (1895-1902). Chancellor Mining Co. (1909). Slate Creek Gold Mining Co. (1915). New Light Mining Co. (1933-1936). Monica Mines, Inc. (1939). Slate Creek Mining Co. (1941-1952). **Ore:** Gold, silver. **Ore min:** Free gold, pyrite, sylvanite. **Deposit:** Fracture zone in limy and graphitic argillite contains interlacing quartz veinlets, graphitic shear zones, and breccia zones cemented by quartz and sulfides. **Dev:** Several thousand ft. on 6 levels. **Improv:** Camp buildings and 100-ton flotation mill (1951). **Assays:** Early work, $30 to $40 Au, later work 0.2 to 0.4 oz. Au, in 1950 about $12 Au. **Prod:** 60,000 tons of ore in the 1900's and several tons of good-grade ore in 1940 to 1942. Small amount 1949. **Ref: 1,** no. 2, 1917, p. 29. **46,** pp. 234-235. **63,** p. 57. **88,** p. 47. **97,** 1907, 1935, 1938-1940. **99,** 2/26/35. **104,** 6/15/32, p. 18; 10/30/34, p. 22; 9/30/35, p. 24; 8/30/36, p. 26; 11/30/36, p. 29. **105,** 1895, p. 106. **106,** 6/4/31; 8/3/33. **114,** no. 2, 1906, p. 22; no. 5, 1909, pp. 88, 90. **150,** p. 39. **158. 159,** p. 133.

Nooksack (3)

Loc: Near E. ¼ cor. sec. 35, (40-4E), Mt. Baker dist. **Elev:** 1,200 ft. **Access:** Trail. **Prop:** 3 properties: Nooksack, Givens-Nooksack, Land-Nooksack. **Owner:** Nooksack Mining Co. (1903). **Ore:** Gold. **Ore min:** Gold-bearing sulfide. **Deposit:** Small quartz stringers in andesite. **Dev:** More than 1,500 ft. of underground workings, mostly caved. **Assays:** It is reliably reported that more than 50 channel samples av. $2 Au at the old price. **Prod:** Probably produced. **Ref: 88,** p. 44. **105,** 4/04, p. 236. **158. 159,** p. 131.

North American (Velvet) (23)

Loc: Sec. 11, (37-16E), between Boulder Cr. and Mill Cr., Slate Cr. dist. **Elev:** 4,300 ft. **Prop:** 10 unpatented claims. **Owner:** North American Mining & Milling Co. (1902-1915). **Ore:** Gold, silver. **Ore min:** Free gold. **Gangue:** Quartz. **Deposit:** 6 veins. **Dev:** 240-ft. adit, 28-ft. adit, 22-ft. adit, 2 other old adits. **Assays:** Av. $15 Au, Ag (1934). **Prod:** Has produced. **Ref: 88,** p. 50. **114,** no. 5, 1909. **116,** no. 10, 1907, p. 18.

Post-Lambert
(See Lone Jack.)

President
(See Great Excelsior.)

Red Mountain
(See Boundary Red Mountain.)

Ruth Mountain Pyrite (13)

Loc: On Ruth Mtn., approx. in secs. 17 and 20, (39-10E), Mt. Baker dist. **Elev:** 2,500 to 6,600 ft. **Access:** Trail from Ruth Cr. road. **Owner:** Abandoned (1949). **Ore:** Gold, iron. **Ore min:** Pyrite. **Deposit:** Said to be 4 veins; one on top of the mountain 30 ft. wide, one halfway down 15 ft. wide, one nearly down 20 ft. wide, and one near the river 10 to 15 ft. wide. **Dev:** Short crosscut adit on 15-ft. vein. **Assays:** Reportedly 28¢ to $176.00 Au. **Ref: 158.**

Saginaw (10)

Loc: W½ sec. 15, (40-9E), Mt. Baker dist. **Access:** 1 mi. by trail from Twin Lakes road. **Prop:** 4 patented claims. **Owner:** Saginaw Gold & Copper Mines, Inc., Bellingham, Wash. (1948 ——). **Ore:** Gold, copper, lead, silver. **Gangue:** Quartz, calcite. **Deposit:** 3½-ft. ledge in diorite. **Dev:** 300-ft. adit with 200 ft. of crosscuts, 76-ft. adit with 31-ft. crosscut, two 36-ft. adits. **Assays:** Picked sample gave 70% Pb, 0.20 oz. Au, 200 oz. Ag. Others gave 0.34 to 1.82 oz. Au, 0.78 to 2.06 oz. Ag, tr. to 2.13% Cu. Av. said to be $6.00 Au, Ag, Cu. **Ref: 88,** pp. 44-45. **150,** p. 38. **158.**

Short Grub

(See Beck and Short Grub.)

Slate Creek

(See New Light.)

Tacoma (Golden Arrow) (21)

Loc: SE¼ sec. 34, (38-17E), on W. side of Slate Cr. **Access:** Road from Mazama. **Prop:** 4 unpatented claims. **Owner:** E. H. Spafford and Walter Gourlie, Twisp, Wash. (1952 ——). Gold Standard Mining Co. (1902). H. C., E. E., and M. J. Blocher (1938). **Ore:** Gold, lead, zinc, silver. **Ore min:** Free gold, tellurides, galena, sphalerite, pyrite, arsenopyrite. **Deposit:** 2 veins, one said to av. 2 ft. thick. One is 3 to 12 in. thick. **Dev:** Shaft, lower adit about 300 ft. long. **Assays:** Tr. to $34.00 Au. **Prod:** 10-ton test shipment netted $60.00 per ton. **Ref: 88,** p. 48. **133,** p. 34. **158.**

Velvet

(See North American.)

Verona (Galena) (12)

Loc: Sec. 20, (39-9E), Mt. Baker dist. **Access:** Near Mt. Baker highway about 4 mi. below the Mt. Baker lodge. **Prop:** 5 claims. **Owner:** Helen Tweit, Bellingham, Wash. (1952). Verona Mining Co. (1934). James Sullivan and Harold Bland, Bellingham, Wash. (1944). **Ore:** Gold, silver, lead, zinc, copper. **Ore min:** Pyrite. **Deposit:** Shear zone in altered volcanic rock contains stringers and lenses of mineralized quartz and calcite. Some stringers as much as 6 in. wide. **Dev:** 65-ft. adit with a 20-ft. drift. **Improv:** Several old cabins. **Assays:** $10 to $20 per ton reported. One assay showed 0.21 oz. Au, 80¢ Ag, 14.5% Pb. **Ref: 158.**

Whistler (24)

Loc: Secs. 8 and 17, (37-17E), Slate Cr. dist. **Access:** Road. **Prop:** 4 patented claims. **Owner:** Marguerite Roth, Bellingham, Wash. (1952). **Ore:** Gold, silver, copper. **Ore min:** Sulfides. **Deposit:** Ore bodies said to be 2 to 6 ft. wide. **Assays:** Reportedly $15 to $20 per ton. One assay reported to show $48 Au, 3.7% Cu. **Prod:** 1936, 1937. **Ref: 63,** p. 58. **97,** 1937, p. 553; 1938, p. 460. **158.**

Willis
(See Lone Jack.)

YAKIMA COUNTY

Bear Gap
(See Fife.)

Blue Bell
(See Fife.)

Crown Point (69)

Loc: A little W. of the Comstock property, Summit dist. **Owner:** George M. Brown (1897). **Ore:** Gold, silver. **Deposit:** 7-ft. vein. **Dev:** 30-ft. adit. **Assays:** Av. $38 Au, Ag (1897). **Ref: 63**, p. 44.

Fife (Bear Gap, Pickhandle, Manitau, Blue Bell) (70)

Loc: Sec. 31, (17-11E), along the Cascade crest from Bear Gap to Crown Point. **Elev:** 5,000 to 6,000 ft. **Access:** Road up Morse Cr. from Chinook Pass highway reaches within ½ mi. of the property. **Prop:** 2 patented claims: Cold Spring, Silver Reef; and 12 unpatented claims. **Owner:** Manitau Mining & Milling Co. (1941). **Ore:** Gold, silver. **Ore min:** Free gold. **Deposit:** Small quartz veinlets in volcanic rock. Volcanic rock itself reputedly carries gold values. **Assays:** 90 channel samples 10 to 20 ft. in length over an area 1,200 by 1,800 ft. showed an av. value of $1.37 Au. **Prod:** 1896. **Ref: 13**, p. 162. **63**, p. 44. **158**.

Gold Hill (71)

Loc: S½ sec. 31, (17-11E), Summit dist. **Elev:** 4,500 to 6,000 ft. **Access:** Morse Cr. road crosses the property. **Prop:** 4 patented claims: Eureka, Climax, Lake Paragon, Boston; and 24 unpatented claims. **Owner:** E. A. Bannister and G. A. Mosher, Yakima, Wash. **Ore:** Gold, reportedly lead, zinc, copper, silver, tungsten. **Ore min:** Free gold. **Deposit:** Small quartz veinlets in volcanic rock. Volcanic rock itself reportedly carries gold values. **Dev:** 2 adits, each about 500 ft. long. **Assays:** Hand samples assayed 75¢ to $1.50 Au. **Ref: 58**, p. 25. **158**.

Manitau
(See Fife.)

Pickhandle
(See Fife.)

Lode Gold Properties

1. Port Angeles
2. Rustler Creek
3. Nooksack
4. Great Excelsior
5. Gargett
6. Gold Run
7. Boundary Red Mountain
8. Evergreen
9. Lone Jack
10. Saginaw
11. Goat Mountain
12. Verona
13. Ruth Mountain Pyrite
14. Anacortes
15. Allen Basin
16. Beck and Short Grub
17. Chancellor
18. Goat
19. Mammoth
20. New Light
21. Tacoma
22. Anoka
23. North American
24. Whister
25. Azurite
26. Soldier Boy
27. Mount Vernon
28. Blue Bird
29. Queen Anne
30. Sunrise
31. Wayside
32. Copper Independent
33. Eclipse
34. "45"
35. Glory of the Mountain
36. Rainy
37. Monte Cristo
38. Caplin-Holbrooke
39. Good Hope
40. Justice
41. Peabody
42. Sidney
43. Ben Butler
44. Jasperson
45. Index Gold Mines, Inc.
46. Last Chance
47. Great Northern
48. Calumet
49. Edison
50. Mineral Center
51. Undaunted
52. Washington-Iowa
53. Damon and Pythias
54. Apex
55. Coney Basin
56. Beaverdale
57. Lucky Strike
58. Monte Carlo
59. Lennox
60. Carmack
61. San Jose
62. White River
63. Seigmund Ranch
64. Silver Creek Gold & Lead
65. Blue Grouse and Sure Thing
66. Campbell
67. Silver Creek
68. Washington Cascade
69. Crown Point
70. Fife
71. Gold Hill
72. Golconda
73. Grizzly Creek
74. Bruhn
75. Camp Creek
76. Johnson
77. Perry
78. Primary Gold
79. Brown and Livingston
80. Plamondon
81. Green Mountain
82. Golden Wonder
83. Silver Star
84. Mountain Beaver

85. Mazama Queen
86. Imperial
87. American Flag
88. Gold Key
89. Mazama Pride
90. Iron Cap and Snow Cap
91. Mid Range
92. Spokane
93. Red Shirt
94. Alder
95. Minnie
96. Silver Ledge
97. Independence
98. Gold Coin
99. St. Anthony
100. Grubstake
101. Holden-Campbell
102. Okanogan
103. Hunter
104. Hidden Treasure
105. Highland
106. Methow
107. Friday
108. Roosevelt
109. Last Chance
110. Paymaster
111. Chelan
112. Hidden Treasure
113. Holden
114. Red Cap
115. Red Hill
116. Butte
117. Kingman and Pershall
118. Pangborn
119. Sunshine
120. Rex
121. Cook-Galbraith
122. Aurora
123. Silver Bull
124. Silver Creek
125. Ida Elmore
126. Maud O.
127. Alta Vista
128. Black and White
129. Black Jack
130. Blewett
131. Blue Bell
132. Bobtail
133. Culver
134. Eureka
135. Fraction
136. Golden Eagle
137. Hummingbird
138. La Rica
139. North Star
140. Olden
141. Olympia
142. Peshastin
143. Phipps
144. Phoenix
145. Pole Pick
146. Prospect
147. Sandell
148. Tip Top
149. Wilder
150. Lucky Queen
151. Blinn
152. Ontario
153. Gold Knob
154. Golden King
155. Golden Fleece
156. Wall Street
157. Cascade Chief
158. Cougar
159. Flodine
160. Liberty
161. Ewell
162. Clarence Jordin
163. Mountain Daisy
164. Ollie Jordin
165. Prosser
166. Golden Zone
167. Gold Crown
168. Chloride Queen
169. Pinnacle
170. Leadville

171. Gold Crown
172. Summit
173. Palmer Mountain Tunnel
174. Bullfrog
175. Okanogan Free Gold
176. Jessie
177. Bellevue
178. Hiawatha
179. Occident
180. Spokane
181. Triune
182. Palmer Summit
183. Rainbow
184. Black Bear
185. War Eagle
186. Alice
187. Poland China
188. Butcher Boy
189. Gray Eagle
190. Reco
191. Caribou
192. Gold Axe
193. Crystal Butte
194. Whitestone
195. Bodie
196. Silver Bell
197. Morning Star
198. Surprise
199. Panama
200. Valley
201. Hawkeye
202. Tom Thumb
203. Rebate
204. South Penn
205. Anecia
206. Ben Hur
207. Black Tail
208. El Caliph
209. Ida May
210. Iron Mask
211. Knob Hill
212. Little Cove
213. Lone Pine
214. Morning Glory
215. Mountain Lion
216. Pearl
217. San Poil
218. Seattle
219. Surprise
220. Trade Dollar
221. Insurgent
222. Last Chance
223. Quilp
224. Advance
225. Butte and Boston
226. Flag Hill
227. Princess Maude
228. Republic
229. California
230. Golden Harvest
231. Rosario
232. White Elephant
233. F. H. and C.
234. Swamp King
235. Beecher
236. Eureka
237. Easter Sunday
238. Titanic
239. First Thought
240. First Thought Extension
241. Michigan
242. Trophy
243. Second Thought
244. Gem
245. Homestake
246. Antelope
247. St. Crispin
248. Gold Bar
249. Gold Ledge
250. Gold Reef
251. Sunday
252. Rocky Creek
253. Deemer
254. Gilbert
255. Sunrise
256. Hansen

Placer Gold Properties

1. Shi Shi Beach
2. Ozette Beach
3. Little Wink
4. Morgan
5. Morrow
6. Yellow Banks
7. Main and Bartnes
8. Johnson Point
9. Cedar Creek
10. Sunset Creek
11. Ruby Beach
12. Moclips River
13. Oyhut
14. Point Brown
15. Ocean Park
16. Fort Canby
17. Sand Island
18. Scougale
19. Jackass
20. Combination
21. Lazy Tar Heel
22. Nip and Tuck
23. Alice Mae
24. Woodrich
25. Farrar
26. Ruby Creek
27. Johnnie S.
28. Old Discovery
29. Anacortes
30. Day Creek
31. Darrington
32. Deer Creek
33. Granite Falls
34. Peterson
35. Alpha and Beta
36. Sultan Canyon
37. Aristo
38. Horseshoe Bend
39. Sultan River
40. Sultan
41. Gold Bar
42. Bench
43. Phoenix
44. Money Creek
45. Tolt River
46. Raging River
47. Elizabeth
48. Gold Links
49. Gold Hill
50. Ogren
51. Silver Creek
52. Hudson and Meyers
53. Surveyors Creek
54. Texas Gulch
55. Lewis River
56. McMunn
57. Brush Prairie
58. Similkameen
59. Similkameen Falls
60. Meadows
61. Ballard
62. Methow River
63. Cassimer Bar
64. Stehekin River
65. Railroad Creek
66. Deep Creek
67. Mad River
68. Entiat River
69. Icicle Creek
70. Leavenworth
71. Wenatchee River
72. Wednesday
73. Wenatchee
74. Ingalls Creek
75. Ruby Creek
76. Bloom
77. Solita
78. Shaser Creek
79. Nigger Creek
80. Fortune Creek
81. Big Salmon La Sac
82. Cle Elum
83. Bear Cat
84. Baker Creek
85. Naneum Creek
86. Boulder Creek
87. Nugget
88. Old Bigney
89. Becker
90. Bryant Bar
91. Dennett
92. Swauk Mining & Dredging
93. Williams Creek
94. Gold Bar
95. Swauk Creek
96. Yakima River
97. Perry
98. A. B. C.
99. Chinaman Bar
100. Artesian Coulee
101. Gone Busted
102. Berrian Island
103. Cuba Line
104. Walker
105. Altoona
106. Deadman Creek
107. Mary Ann Creek
108. Goosmus Creek
109. Marcella
110. Alva Stout
111. Crounse
112. Rico
113. Gold Bar
114. Reed and Roberts
115. Nigger Creek Bar
116. Northport Bar
117. Nigger Bar
118. Evans
119. Schierding
120. Harvey Bar
121. Sullivan Creek
122. Schultz
123. Kettle River
124. China Bend
125. Ninemile Bar
126. Bossburg Bar
127. Valbush Bar
128. Brod-Hurst
129. Marcus
130. Holsten
131. Daisy
132. Collins
133. Johnson
134. Stranger Creek
135. Turtle Rapids
136. Blue Bar Island
137. Blue Bar
138. Rogers Bar
139. Wilmont Bar
140. Gibson Bar
141. Ninemile
142. Covington Bar
143. Bacon Bar
144. Hellgate Bar
145. Plum Bar
146. Barnell
147. Clark
148. Keller Ferry
149. Winkelman Bar
150. Creston Ferry
151. Kirby Bar
152. Peach Bar
153. Indian Bar
154. Clarkston
155. Snake River

PLACER GOLD OCCURRENCES

ASOTIN COUNTY

Clarkston Placer (154)

Loc: Snake R. near Clarkston. **Ore:** Gold. **Ore min:** Gold, magnetite, ilmenite, zircon. **Assays:** River sand concentrated by panning gave $24.99 per ton in gold, 572 lb. magnetite, 530 lb. ilmenite, and 30 lb. zircon per ton. **Ref: 126.**

Snake River Placer (155)

Loc: Snake R., Asotin County. **Ore:** Gold. **Ore min:** Gold, zircon, ilmenite, magnetite. **Assays:** Natural sand 12¢ to $1.64 per ton gold, tr. to 2 lb. per ton zircon, 12 to 18 lb. per ton ilmenite, 20 to 34 lb. per ton magnetite. **Ref: 38-A,** p. 1216. **126,** p. 14.

BENTON COUNTY

Artesian Coulee Placer (100)

Loc: Sec. 6, (4-24E), 4 mi. E. of Alderdale. **Elev:** 250 ft. **Access:** On railroad. **Ore:** Gold. **Deposit:** Gravel in a deserted channel of the Columbia R. covering several square miles. **Assays:** 50¢ to $2.00 per cu. yd. **Ref: 158.**

Berrian Island (Goody) Placer (102)

Loc: Sec. 1, (5-28E) and sec. 6, (5-29E), on N. bank of Columbia R. **Elev:** 300 ft. **Owner:** A. H. Goody, Inc., Tacoma, Wash. (1950). **Ore:** Flour gold. **Deposit:** Screened material from the J. G. Shotwell aggregate plant was run through sluice boxes and a low-grade conc. recovered. **Assays:** 2 to 6 mills per ton. **Prod:** Plant operated 6 weeks and conc. shipped to Tacoma smelter in 1949. **Ref. 158.**

Blalock Island Placer

Loc: On Blalock Is., in Columbia R., near Patterson. **Prop:** State leases. **Owner:** Lessees in 1955 are R. C. Matney, Prosser, Wash., Fred Blake, Grandview, Wash., and others. **Ore:** Gold. **Prod:** Small amount in 1954 and unknown amount in previous years. **Ref: 69-A,** p. 8. **158.**

Gone Busted Placer (101)

Loc: On Blalock Is., in Columbia R., near Patterson. **Ore:** Gold. **Prod:** Dry-land washing plant operated 1938-1940. **Ref: 97,** 1937, p. 458; 1938, p. 489; 1939, p. 476; 1940, p. 472.

Goody Placer
(See Berrian Island Placer.)

CHELAN COUNTY

Bloom Placer (76)

Loc: Sec. 1, (22-17E), on Peshastin Cr., 1 mi. above Nigger Cr., Blewett dist. **Elev:** 2,200 ft. **Access:** Near U. S. Highway 97. **Prop:** 3 claims. **Owner:** G. W. and J. M. Bloom and John Snider (1897). **Ore:** Gold. **Assays:** About 25¢ per yd. **Prod:** About $100 by 1897. **Ref: 63**, p. 77. **67**, p. 48.

Deep Creek Placer (66)

Loc: Sec. 19, (27-18E), ·at the mouth of Deep Cr. **Prop:** 13 claims. **Owner:** Deep Creek Mining Co. (1897). **Ore:** Gold. **Assays:** About 26¢ per yd. **Prod:** Unknown amount. **Ref: 63**, p. 79. **67**, p. 48.

Entiat River Placers (68)

Loc: Along Entiat R. **Ore:** Gold. **Ref: 13**, p. 174.

Icicle Creek Placers (69)

Loc: Along Icicle Cr. **Ore:** Gold. **Prod:** Reportedly considerable. **Ref: 13**, p. 174.

Ingalls Creek Placer (74)

Loc: Sec. 25, (23-17E), on Peshastin Cr. at the mouth of Ingalls Cr. **Owner:** Mr. Hensel (1897). **Ore:** Gold. **Prod:** Unknown amount. **Ref: 67**, p. 48.

Leavenworth Placer (70)

Loc: Secs. 10 and 11, (24-17E), near Leavenworth. **Ore:** Gold. **Ref: 67**, p. 48.

Mad River Placer (67)

Loc: On Mad R. **Ore:** Gold. **Ref: 105**, 1898, p. 613.

Nigger (Negro) Creek Placers (79)

Loc: Secs. 2 and 3, (22-17E), from mouth of Nigger Cr. upstream for 2 mi. **Prop:** Many claims. **Owner:** Several. **Ore:** Gold. **Assays:** 10¢ to $1.20 per yd. **Prod:** $1,100 prior to 1897. **Ref: 67**, p. 48.

Peshastin Creek Placers

Loc: On upper reaches of Peshastin Cr. **Ore:** Coarse gold. **Deposit:** Placer gold found in best concentration on bedrock below gravels. **Assays:** Gold is of high purity. **Prod:** Has produced. **Ref: 12**, p. 8. **13**, p. 173. **143**, p. 78. **144**, p. 9.

Railroad Creek Placer (65)

Loc: Secs. 16 and 17, (31-18E), on Railroad Cr. **Ore:** Gold. **Ref: 67**, p. 48.

Ruby Creek Placer (75)

Loc: Sec. 36, (23-17E), at mouth of Ruby Cr. **Prop:** 6 claims. **Owner:** James and Thomas Lynch, Riley Eisenhour, and Thomas Medhurst (1897). **Ore:** Gold. **Ref: 67,** p. 48.

Shaser Creek Placer (78)

Loc: SE¼ sec. 14, (22-17E), near mouth of Shaser Cr. **Ore:** Gold. **Assays:** Gold is of high purity. **Prod:** Unknown amount. **Ref: 67,** p. 48. **144,** p. 9.

Solita Placer (77)

Loc: On Peshastin Cr. **Ore:** Gold. **Prod:** Unknown amount in 1931. **Ref: 67,** p. 49.

Stehekin River Placers (64)

Loc: Along Stehekin R. **Ore:** Gold. **Ref: 13,** p. 174.

Wednesday Placer (72)

Loc: On Wenatchee R. near Dryden. **Ore:** Gold. **Prod:** Unknown amount in 1931. **Ref: 67,** p. 49.

Wenatchee Placer (73)

Loc: Sec. 3, (22-20E), at Wenatchee. **Ore:** Gold. **Ref: 67,** p. 49.

Wenatchee River Placer (71)

Loc: Sec. 22, (24-18E), on Wenatchee R. 1½ mi. below Peshastin. **Owner:** W. M. Keene and O. A. Benjamin, Seattle, Wash. (1897). **Ore:** Gold. **Assays:** As much as $1.00 per yd. **Prod:** Unknown amount. **Ref: 67,** p. 49.

CLALLAM COUNTY

Bartnes Placer

(See Main and Bartnes Placer.)

Big Wink Creek Placer

(See Morgan Placer.)

Cedar Creek (Starbuck) Placer (9)

Loc: Near mouth of Cedar Cr. in E½ sec. 18, (29-15W). **Elev:** Sea level. **Access:** Up the beach 10 mi. N. of La Push. **Owner:** J. M. Starbuck (1917). **Ore:** Gold, platinum. **Ore min:** Free gold, platinum, ilmenite, magnetite, chromite, zircon. **Deposit:** Gold and platinum occur in a beach placer in a 2- to 15-in. sand layer on the surface of a wave-cut bench in clay. The sand layer is said to be 400 ft. long and 100 ft. wide at low tide. **Assays:** Said to yield from 2 to 5 pennyweights of Au and Pt per cu. yd. Ratio of Pt to Au ranges from 1:5 to 1:15. **Prod:** Reported $5,000 in gold and 5 oz. of crude platinum prior to 1917. **Ref: 126,** pp. 3-5.

Japanese Placer
(See Little Wink Placer.)

Johnson Point Placer (8)

Loc: NW¼ sec. 5, (28-15W), at Johnson Point (Cape Johnson). **Access:** Reached by hiking 4 mi. N. along beach from La Push. **Ore:** Gold. **Deposit:** Beach placers and gold in Pleistocene deposits of the bench adjacent. **Assays:** Some pans of sand ran as high as 60¢ Au, and some panners recovered $2 to $10 per day. **Prod:** 1908. **Ref: 135-A**, p. 233.

Little Wink (Japanese, Sand Point) Placer (3)

Loc: Center S½SW¼ sec. 1, (30-16W), at mouth of Little Wink Cr. **Access:** Trail. **Ore:** Gold. **Deposit:** Reworked Pleistocene gravels in a beach placer. **Ref: 158.**

Lovelace Placer
(See Shi Shi Beach Placer.)

Main and Bartnes Placer (7)

Loc: NW¼NW¼ sec. 19, (30-15W), near small stream. **Access:** Trail. **Prop:** 1 placer claim. **Owner:** Lee Main and A. Bartnes, Everett, Wash. (1940). **Ore:** Gold. **Deposit:** Sand and gravel cemented by clay and iron oxide. Some boulders a foot in dia. **Dev:** Irregular pit 30 ft. long, 15 ft. wide, and 6 ft. deep. **Improv:** Flume 150 ft. long (1940). **Ref: 158.**

Morgan (Big Wink Creek) Placer (4)

Loc: NW¼NW¼NE¼ sec. 12, (30-16W), on beach 125 ft. NW. of mouth of Big Wink Cr. **Access:** Trail. **Owner:** S. D. Morgan (1940). **Ore:** Flour gold. **Deposit:** Reworked Pleistocene gravels at high-tide level. Boulders as much as 2 ft. in dia.; those 1 ft. in dia. are common. **Dev:** Pit 30 ft. long, 15 ft. wide, and 6 ft. deep. **Ref: 158.**

Morrow Placer (5)

Loc: SW¼SW¼ sec. 18, (30-15W), at high-tide level. **Elev:** 5 ft. **Access:** Trail. **Prop:** Placer claims. **Owner:** J. D. Morrow (1940). **Ore:** Flour gold. **Deposit:** Beach placer in which there is a 2-ft. layer of gravel reworked by wave action from Pleistocene gravel. **Dev:** An area 50 by 25 ft. has been worked. **Assays:** Spotty fine gold content av. $4.00 per yd. **Prod:** Worked from 1932 to 1940. Best year's cleanup was $1,678. **Ref: 158.**

Ozette Beach Placer (2)

Loc: Sec. 12, (31-16W), on the beach 2 mi. N. of mouth of Ozette R. **Elev:** Sea level. **Access:** Trail from Lk. Ozette. **Ore:** Gold, platinum. **Deposit:** Beach deposit of gold and platinum concentrated along surface of a wave-cut terrace in sandstone.

Prod: Small amount in early 1900's. **Ref: 3,** p. 156. **97,** 1916, p. 610. **126,** pp. 3, 5. **135-A,** pp. 232-233.

Sand Point Placer
(See Little Wink Placer.)

Shi Shi Beach (Lovelace) Placer (1)

Loc: Beach between Portage Head and Point of Arches in secs. 18, 19, and 30, (32-15W). **Elev:** Sea level. **Access:** Road from Neah Bay reaches place on cliff above the beach. **Owner:** W. W. Lovelace (1908). **Ore:** Gold, platinum, iridium, osmium. **Ore min:** Magnetite, ilmenite, free gold, platinum, iridosmine, zircon. **Deposit:** Beach deposit in which a thin layer of fine heavy sand rests on a wave-cut terrace in sandstone overlain by 1 to 3 ft. of gravel and sand. Gold and platinum are found in the thin layer and in cracks in the bedrock. **Assays:** A conc. showed 1,120 lb. ilmenite, 96 lb. zircon, $558.09 gold, $20.45 platinum per ton. **Prod:** $15,000 in gold prior to 1904 from this and the Ozette placer. **Ref: 3,** pp. 154-156. **38-A,** pp. 1218-1219. **47,** vol. 17, 1906, p. 467. **97,** 1908, p. 578. **126,** p. 14. **135-A,** p. 232.

Starbuck Placer
(See Cedar Creek Placer.)

Sunset Creek Placer (10)

Loc: Sec. 19, (29-15W). **Ore:** Gold, platinum. **Ref: 158.**

Yellow Banks Placer (6)

Loc: 2 mi. S. of Sand Point in SW¼ sec. 18, (30-15W). **Elev:** Sea level. **Access:** Reached by hiking S. along beach from the Ozette R. or N. from La Push. **Owner:** D. J. Wright (1917). **Ore:** Gold, platinum. **Ore min:** Free gold, platinum, ilmenite, magnetite. **Deposit:** Beach deposit of gold and platinum concentrated on surface of a wave-cut terrace in Pleistocene drift. **Prod:** Small amount in early 1900 and in 1939. **Ref: 3,** pp. 156-157. **47,** vol. 17, p. 467. **97,** 1940, p. 476. **126. 135-A,** p. 233. **141,** p. 103.

CLARK COUNTY

Brush Prairie Placer (57)

Loc: On Brush Prairie. **Ore:** Gold. **Ore min:** Free gold, magnetite, ilmenite. **Deposit:** Placer "black sands." **Assays:** A sample of conc. showed 1,176 lb. magnetite, 328 lb. ilmenite, $57.05 gold per ton (1905). **Ref: 38-A,** pp. 1218-1219. **126,** p. 14.

Lewis River Placer (55)

Loc: 5 mi. E. of Moulton on the Lewis R., 1 mi. above McMunn placer. **Elev:** 850 ft. **Access:** Road up E. Fk. Lewis R. **Owner:** Company in Portland, Oregon (1912). **Ore:** Gold. **Deposit:** Placer gravel in river bed. **Dev:** Pit 75 by 100 ft. and 30 ft. deep.

Improv: $30,000 hydraulic mining outfit (1912). **Assays:** 1 oz. Au recovered from 1 ton of black sand. **Prod:** 1 oz. of gold. **Ref: 126,** pp. 6-7.

McMunn Placer (56)

Loc: On E. Fk. Lewis R. about 4 mi. E. of Moulton. Approx. sec. 21, (4-4E). **Elev:** 800 ft. **Access:** Road up E. Fk. Lewis R. **Prop:** 1 placer claim. **Owner:** Harvey McMunn (1912). **Ore:** Gold, platinum. **Ore min:** Free gold, platinum, magnetite, hematite, limonite. **Deposit:** Gold and platinum occur in a thin veneer of gravel on bedrock and in cracks in the bedrock. Gold but not platinum found also in higher terrace gravels. **Assays:** Panning tests indicate the pay gravel to contain 0.126 gm. Pt, 1.14 gm. Au. **Prod:** McMunn reports $150 in gold and 1½ oz. platinum. **Ref: 126,** pp. 6-9. **130,** p. 85. **141,** p. 103.

DOUGLAS COUNTY

Columbia River Placer

Loc: Columbia R. **Ore:** Gold. **Ore min:** Gold, magnetite, chromite, ilmenite, zircon, monazite. **Assays:** 13 lb. of black sand recovered from 1 cu. yd. of gravel showed 1,414 lb. magnetite, 150 lb. chromite, 188 lb. ilmenite, 24 lb. zircon, 6 lb. monazite, $30.59 gold per ton (1905). **Ref: 38-A,** pp. 1218-1219. **126,** p. 14.

Trouble Placer

Loc: On Columbia R. at Stevenson's Ferry landmark. **Prop:** 1 placer claim. **Owner:** Mrs. Rebecca Stevenson, Barry, Wash. (1938). **Ore:** Gold. **Assays:** One man panned $105 in 5 days. **Ref: 158.**

FERRY COUNTY

Alva Stout Placer (110)

Loc: NW¼ sec. 2, (36-32E), Republic dist. **Ore:** Gold. **Prod:** 1934. **Ref: 97,** 1935, p. 353.

Bacon Bar Placer (143)

Loc: N½ sec. 15, (28-33E), on N. side of Columbia R., about 1 mi. upstream from mouth of Sanpoil R. **Ore:** Gold. **Assays:** 15 samples showed 8¢ to 13¢ per cu. yd. **Ref: 158.**

Blue Bar Island Placer (136)

Loc: Bar in Columbia R. near mouth of Stray Dog Cr., sec. 12, (31-36E). **Ore:** Flour gold. **Deposit:** Flour gold occurs in surficial layer of gravel in low bar. Said to be renewed by annual freshets. **Assays:** Panning tests indicate that a miner with sluice or rocker might wash out 50¢ worth per day. **Ref: 26,** p. 66. **122,** p. 177.

Covington Bar Placer (142)

Loc: Sec. 9, (28-33E), on N. side of Columbia R. **Owner:** Henry Covington (1934). **Ore:** Gold. **Assays:** 19 test holes showed no values of importance except in a small area of less than 1 acre, where 5 holes tested gave an av. of 30¢ per cu. yd. **Ref:** **158.**

Daisy Placer (131)

Loc: On W. bank of Columbia R., 2 mi. above Daisy. Approx. in sec. 1, (33-37E). **Ore:** Gold. **Deposit:** Terraces 20 to 30 ft. above river level contain a 6-in. paystreak at a depth of 1 ft. **Assays:** Samples of the paystreak ran 2¢ or 3¢ per cu. yd. **Ref: 26,** p. 68.

Goosmus Creek Placer (108)

Loc: On Big Goosmus Cr., Danville area. **Owner:** A. E. Milton and Roy Hathaway, Curlew, Wash. (1950). **Ore:** Gold. **Deposit:** Placer deposit containing many boulders, along narrow creek bottom. **Improv:** Small sluicing outfit (1949). **Ref: 69,** p. 9. **158.**

Hellgate Bar Placer (144)

Loc: Lot 5, sec. 13, (28-33E), N. side of Columbia R. **Owner:** United States. **Ore:** Gold. **Deposit:** A high bar had gravel 2½ to 3 ft. thick showing values of 21.8¢ per cu. yd. A lower bar had 7 or 8 acres of gravel 2 to 3 ft. thick av. 27¢ per cu. yd. A low bench 2 mi. long and 300 ft. to ½ mi. wide contains an iron-stained gold-bearing layer about 6 in. thick, 100 to 600 yd. wide, and 2 mi. long. **Assays:** The 6-in. paystreak av. 1¢ to 5¢ per cu. yd. **Prod:** Small amount in early 1900's. **Ref: 26,** pp. 62-63. **158.**

Johnson Placer (133)

Loc: On bank of Columbia R. in NE¼ sec. 8, (32-37E). **Prop:** 1 placer claim. **Owner:** W. Johnson. **Ore:** Gold. **Deposit:** Paystreak consists of 1 to 3 ft. of medium- to fine-textured gravel beneath 4 to 8 ft. of sand. **Dev:** Pit 2,000 cu. yd. in size. **Assays:** Av. of 5¢ per yd. from the pit. **Prod:** $100 reported. **Ref: 122,** pp. 177-178.

Kettle River Placers (123)

Loc: Along Kettle R. **Ore:** Gold. **Ref: 13,** p. 174.

Kirby Bar Placer (151)

Loc: Secs. 15 and 16, (27-35E), on N. side of Columbia R., opposite Peach. **Owner:** J. M. Kirby, Peach, Wash. (1934). **Ore:** Gold. **Assays:** Panning indicated 38.7¢ per cu. yd. **Ref: 158.**

Marcella Placer (109)

Loc: On Granite Cr., Republic dist. **Ore:** Gold. **Prod:** 1939. **Ref: 97,** 1940, p. 477.

Ninemile Placer (141)

Loc: On Columbia R. near mouth of Ninemile Cr., probably in sec. 16, (29-35E). **Ore:** Gold. **Ore min:** Flake gold worth 0.02¢ per flake and much flour gold. **Deposit:** 2 terraces, one 30 ft. and the other 100 ft. above Columbia R. Paystreaks as much as 1 to 3 ft. thick in the terraces. **Dev:** Old Chinese workings extend ¾ mi. along the river and 200 ft. back from the river. **Assays:** Paystreak av. a little more than 40¢ per yd. **Ref: 26,** pp. 64-65.

Plum Bar Placer (145)

Loc: Lots 7 and 8, sec. 17, (28-32E), about 6 mi. up the Columbia R. from Grand Coulee Dam, on N. side of the river. **Owner:** Col. White (1934). **Ore:** Gold. **Assays:** 77,000 cu yd. av. 33.1¢ per yd. **Prod:** J. H. Collins washed 700 to 800 cu. yd. per day in 1934. Produced 1938-1939. **Ref: 97,** 1939, p. 490; 1940, p. 477. **104,** 6/15/35, p. 27. **158.**

Rico Placer (112)

Loc: 20 mi. S. of Republic. **Ore:** Gold. **Assays:** 50¢ to 75¢ per cu. yd. **Ref: 105,** 1911, p. 417.

Rogers Bar Placer (138)

Loc: Sec. 23, (30-36E), on W. bank of Columbia R., 2 mi. below the town of Hunters. **Prop:** Bar is about 3½ mi. long. **Owner:** Colville Indian Reservation. **Ore:** Gold, platinum. **Deposit:** 3 bars, 30, 75, and 100 ft. above the river, include 1,500 acres of land. The best pay gravel is found in bars exposed only at low water. **Assays:** Tests indicate paystreaks in the lower bar to run 1¢ to 50¢ per yd. in gold. **Prod:** 4 operators were working in 1934. **Ref: 26,** pp. 65-66. **97,** 1935, p. 352. **158.**

Stout Placer

(See Alva Stout Placer.)

Stranger Creek Placer (134)

Loc: At mouth of Stranger Cr., opposite Gifford. **Ore:** Gold. **Deposit:** Bar contains a 1- to 3-ft. paystreak at a depth of 7 ft. Sampling indicates the paystreak to have a very limited extent. **Assays:** Panning tests indicate the paystreak runs 5½¢ to 40¢ per cu. yd. **Ref: 26,** p. 67.

Turtle Rapids Placer (135)

Loc: On the Columbia R. 2 mi. above Blue Bar, near Turtle Rapids, near Covada. **Ore:** Gold. **Deposit:** Terrace 60 ft. above high water extending along the river for several miles contains a paystreak a few inches thick. **Dev:** Worked by Chinese. **Assays:** Panning tests indicate the paystreak to run 30¢ per cu. yd. **Ref: 26,** pp. 66-67.

Wilmont Bar Placer (139)

Loc: Lot 6, sec. 4, (29-36E), on N. bank of Columbia R., opposite Gerome post office. **Ore:** Gold, cerium, thorium. **Ore min:**

Monazite, magnetite, ilmenite, zircon, free gold. **Deposit:** 2 terraces, one 20 and the other 100 ft. above Columbia R. In lower terrace gold occurs in surficial 1 to 5 ft. of material. **Assays:** Lb. per ton: magnetite, 1,308; ilmenite, 150; monazite, 30; zircon, 60; gold, $1.65 per ton. Panning tests indicate a value of 10¢ to 14¢ gold per cu. yd. in lower bar and a fraction of a cent per cu. yd. in the upper bar. **Prod:** 1934. **Ref:** **26,** p. 65. **38-A,** pp. 1218-1219. **97,** 1935, p. 352. **126,** p. 15. **158.**

GRANT COUNTY

A. B. C. Placer (98)

Loc: Sec. 22, (18-23E). **Ore:** Gold. **Ref: 158.**

Chinaman Bar Placer (99)

Loc: Secs. 10 and 11, (13-24E), Chinaman Bar, on Columbia R., 4 mi. E. of Priest Rapids. **Owner:** A. H. and J. H. Miller, Ellensburg, Wash. (1941). **Ore:** Gold. **Prod:** 1939-1941. **Ref:** **22,** p. 8. **97,** 1940, p. 477; 1941, p. 473.

GRAYS HARBOR COUNTY

Cow Point Placer

Loc: On ocean beach, at Cow Point. **Ore:** Gold. **Deposit:** Ocean beach placer. **Assays:** A large sample showed 0.02 oz. Au per ton. **Ref: 38-A,** pp. 1218-1219. **126,** pp. 13-14.

Moclips River Placer (12)

Loc: Sec. 8, (20-12W). **Ore:** Gold. **Ref: 158.**

Oyhut Placer (13)

Loc: On ocean beach at Oyhut. **Ore:** Gold. **Deposit:** Beach placer. **Assays:** 72¢ Au per ton. **Ref: 38-A,** pp. 1216-1217. **126,** p. 14.

Point Brown Placer (14)

Loc: Secs. 15 and 23, (17-12W). **Ore:** Gold. **Ref: 158.**

JEFFERSON COUNTY

Ruby Beach Placer (11)

Loc: E½NE¼ sec. 31, (26-13W), on tombolo between Abbey Is. and Ruby Beach. **Access:** Road. **Ore:** Gold. **Deposit:** Very fine grained gold in beach sand. **Improv:** Gold recovery plant built about 1916, but never operated. **Ref: 158.**

KING COUNTY

Money Creek Placer (44)

Loc: Secs. 20 and 29, (26-11E). **Ore:** Gold. **Ref: 158.**

Raging River Placers (46)

Loc: Along Raging R. **Ore:** Gold. **Ref: 13,** p. 172.

Snoqualmie River Placers

Loc: Along Snoqualmie R. **Ore:** Gold. **Ref: 13,** p. 172.

Tolt River Placer (45)

Loc: SE¼ sec. 29, (26-8E), on Tolt R. **Ore:** Gold. **Deposit:** Fine placer gold said to occur in top 18 in. of gravel in the river bars in this vicinity. **Ref: 158.**

KITTITAS COUNTY

Baker Creek Placers (84)

Loc: Along Baker Cr. above its junction with Swauk Cr. **Ore:** Gold. **Deposit:** Placers. One nugget found on a bench of Swauk Cr. above the mouth of Baker Cr. had a value of $1,004. **Ref: 88,** p. 88.

Bear Cat Placer (83)

Loc: Sec. 33, (21-17E), Swauk dist. **Ore:** Gold. **Prod:** 1930. **Ref: 97,** 1930, p. 673.

Becker Placer (89)

Loc: Sec. 10, (20-17E), Swauk dist. **Ore:** Gold. **Prod:** Has produced. **Ref: 97,** 1915, p. 570; 1916, p. 612.

Big Salmon La Sac Placer (81)

Loc: On Cle Elum R. at Big Salmon La Sac. **Owner:** R. DeWitt and William Taylor (1897). **Ore:** Gold. **Ref: 63,** p. 66.

Boulder Creek Placer (86)

Loc: At junction of Boulder and Williams Creeks, sec. 1, (20-17E). **Owner:** Salem Mining Co. (1936). **Ore:** Gold. **Deposit:** Stream gravels. Pay dirt found at or closely above bedrock. **Assays:** 50¢ to $40 per yd. **Prod:** Considerable. (1936). **Ref: 97,** 1937, p. 551. **143,** p. 77. **144,** p. 9.

Bryant Bar (Deer Gulch) Placer (90)

Loc: Sec. 10, (20-17E), Swauk dist. **Prop:** 65 acres. **Owner:** Deer Gulch Placers, Inc., Seattle, Wash. (1938). **Ore:** Gold. **Ore min:** Coarse free gold. **Deposit:** Cemented gravel on a bench. Bedrock is flat. **Dev:** 325 ft. of drifts. **Assays:** Av. 35¢ per yd., max. 90¢ per yd. **Prod:** 1939. **Ref: 97,** 1940, p. 478. **104,** 2/15/39; 3/15/39. **108,** 12/39, p. 19.

Cle Elum Placer (82)

Loc: On Cle Elum R. near town of Cle Elum. **Ore:** Gold. **Prod:** Reportedly considerable. **Ref: 12,** pp. 9, 95.

Cle Elum River Placer

Loc: Along Cle Elum R. from near headwaters to a place about halfway to its mouth. **Ore:** Gold. **Ref: 13,** p. 173. **63,** p. 66.

Deer Gulch Placer
(See Bryant Bar Placer.)

Dennett Placer (91)
Loc: SE¼ sec. 10, (20-17E), Swauk dist. **Owner:** Cascade Chief Mining Co. (1934). **Ore:** Gold. **Prod:** 1934. **Ref:** 97, 1935, p. 353. 104, 4/30/34, p. 22.

Elliott Placer
Loc: On Williams Cr. **Prop:** 1 claim. **Ore:** Gold. **Assays:** One nugget found in 1900 had a value of $1,100. **Ref:** 63, p. 68. 88, p. 88.

Fortune Creek Placer (80)
Loc: On Cle Elum R. near mouth of Fortune Cr. **Owner:** Messrs. Hicks and Jones (1897). **Ore:** Gold. **Ref:** 63, p. 66.

Gold Bar Placer (94)
Loc: Sec. 15, (20-17E), Swauk dist. **Prop:** 3 claims. **Owner:** B & B Gold Mines, Inc., Cle Elum, Wash. **Ore:** Gold. **Prod:** 1939. **Ref:** 97, 1940, p. 478. 158.

Naneum Creek Placer (85)
Loc: Secs. 25 and 26, (21-18E). **Ore:** Gold. **Ref:** 158.

Nugget Placer (87)
Loc: Secs. 1 and 2, (20-17E) and sec. 6, (20-18E). **Owner:** Nugget Properties, Inc., Seattle, Wash. (1952 ——). **Ore:** Gold. **Prod:** 1952. **Ref:** 108, 1/53, p. 90. 133, p. 37.

Old Bigney Placer (88)
Loc: SW¼ sec. 1, (20-17E), near Liberty. **Ore:** Gold. **Prod:** More than $200,000 prior to 1903. Produced 1908, 1915, 1916, 1923. **Ref:** 97, 1908, 1915, 1916, 1923. 105, 1914, p. 473. 141, p. 22.

Perry Placer (97)
Loc: On Yakima R. about 2 mi. below mouth of Swauk Cr. Probably in sec. 28, (19-17E). **Owner:** F. C. Porter, Thorp, Wash. (1934). **Ore:** Gold. **Dev:** Small pit. **Assays:** 36¢ per yd. **Prod:** Small amount in 1934. **Ref:** 158.

Swauk Creek Placers (95)
Loc: Along Swauk Cr. between the mouths of Baker and First Creeks. **Ore:** Gold. **Deposit:** Gravels from a few feet to 70 or 80 ft. in thickness. Pay gravels are found on and near bedrock. **Assays:** Gravel varies from a few cents to $40 per yd. **Prod:** Considerable. **Ref:** 13, p. 174. 88, pp. 87-88. 143, pp. 76-78. 144, p. 9. 159, pp. 137-138.

Swauk Mining & Dredging Placer (92)
Loc: Secs. 3 and 10, (20-17E). **Owner:** Swauk Mining & Dredging Co., Cle Elum, Wash. (1921-1949). **Ore:** Gold. **Assays:**

Tests prior to 1923 showed 32¢ per yd. **Prod:** 1933. **Ref: 12,** p. 8. **68,** p. 17. **97,** 1921, p. 426; 1934, p. 296. **98,** 1922, p. 1667; 1925, p. 1834; 1926, p. 1597. **129,** p. 276. **141,** p. 22.

Williams Creek Placers (93)

Loc: Along Williams Cr. near Liberty and at its junction with Swauk Cr. **Ore:** Gold. **Deposit:** Good pay gravel is found within 3 or 4 ft. of bedrock and 70 or 80 ft. below present stream level. **Prod:** Considerable. **Ref: 88,** p. 88. **143,** p. 77. **144,** p. 9.

Yakima River Placer (96)

Loc: Secs. 20 and 21, (19-17E). **Ore:** Gold. **Ref: 158.**

LINCOLN COUNTY

Angle Placer

(See Keller Ferry Placer.)

Barnell Placer (146)

Loc: Sec. 7, (28-33E), on the Columbia R. at Swawilla Basin, near Plum, ½ mi. below the ferry. **Owner:** Mr. Mays, Coeur d'Alene, Idaho. **Ore:** Gold. **Prod:** $200 to $400 per week in 1938. **Ref: 158.**

Clark Placer (147)

Loc: SE¼NE¼ sec. 8, (28-33E), along Columbia R. **Ore:** Gold. **Prod:** $4,657 in 1933, $8,243 from 19,700 yd. of gravel in 1934. **Ref: 97,** 1933, p. 296; 1935, p. 353.

Creston Ferry Placer (150)

Loc: Sec. 2, (27-34E), at Creston Ferry. **Ore:** Gold. **Deposit:** River bench. **Dev:** A portion of the bench was worked by placer miners prior to 1910. **Ref: 26,** p. 63.

Keller Ferry (Angle) Placer (148)

Loc: E½ sec. 8, (28-33E), opposite the mouth of Sanpoil R. **Owner:** James E. Angle (1932-1934). **Ore:** Gold. **Prod:** 1932. 186.8 oz. Au from 11,628 cu. yd. of gravel in 1933-1934. **Ref: 104,** 6/30/33, p. 18. **158.**

Peach Bar Placer (152)

Loc: Sec. 22, (27-35E), at village of Peach. **Ore:** Gold. **Deposit:** Low bar. **Assays:** Panning gave an av. of 19¢ per cu. yd. **Ref: 158.**

Winkelman Bar Placer (149)

Loc: S½ sec. 9, (28-33E). **Owner:** Mr. Winkelman (1934). **Ore:** Gold. **Assays:** 2 composite samples showed 22.5¢ and 20.0¢ per cu. yd. **Ref: 158.**

OKANOGAN COUNTY

Altoona Placer (105)

Loc: Myers Cr. dist. **Ore:** Gold. **Prod:** 1902. **Ref:** 100, no. 4, 1902, p. 77.

Ballard Placer (61)

Loc: 1 mi. below Conconully, on Salmon R. **Owner:** Charles H. Ballard and J. R. Wallace (1897). **Ore:** Gold. **Deposit:** Bench about 1 mi. square. **Assays:** 0.1¢ to 10¢ per pan. **Ref:** 63, p. 94.

Cassimer Bar Placer (63)

Loc: At mouth of Okanogan R. **Ore:** Gold. **Prod:** Considerable production reported from 1860 to 1890. **Ref:** 12, pp. 10, 94.

Crounse (Strawberry Creek) Placer (111)

Loc: On Strawberry Cr. in S½ sec. 35, (34-31E). **Access:** 1½ mi. N. of Park City by road. **Owner:** Harry Crounse (1913). **Ore:** Gold. **Ore min:** Free gold, magnetite, ilmenite. **Deposit:** Flats from 20 to 200 ft. wide along the stream are underlain by a shallow layer of coarse gravel. **Dev:** Small pits. **Assays:** 2 pans of gravel from the layer next to bedrock yielded 1¢ in Au and 1 oz. or more of black sand. **Prod:** Reportedly $100 worth of gold. **Ref:** 122, pp. 93, 102-103.

Cuba Line Placer (103)

Loc: Sec. 1, (40-29E), Myers Cr. dist. **Ore:** Gold. **Ref:** 158.

Deadman Creek Placer (106)

Loc: On Deadman Cr., Myers Cr. dist. **Ore:** Gold. **Deposit:** Placers in creek bed and in bars as much as 250 ft. above the creek. **Assays:** As much as 40¢ per yd. **Ref:** 63, p. 110.

Gold Bar Placer (113)

Loc: Near Kartar, along Columbia R. **Owner:** Gold Bar Mining Co. (1939). **Ore:** Gold. **Prod:** 1934, 1939. Washing plant treated 10,045 cu. yd. in 1939. **Ref:** 97, 1935, p. 353; 1940, p. 478.

Mary Ann Creek Placer (107)

Loc: On Mary Ann Cr., 1 mi. S. of Chesaw. **Prop:** 14 claims. **Ore:** Gold. **Deposit:** Gold occurs from grass roots downward to bedrock. 7-in. clay seam 4 ft. above bedrock acts as false bedrock, and values are richer above the clay than elsewhere. **Prod:** $40,000 in 1880's. **Ref:** 88, pp. 27-28. 105, no. 1, 1905, p. 15. 158.

Meadows Placer (60)

Loc: 8 mi. above Conconully, on N. Fk. of Salmon R. **Ore:** Gold. **Deposit:** Bar placer. **Assays:** 10¢ to 15¢ per yd. **Ref:** 63, p. 94.

Methow River Placers (62)

Loc: Along Methow R. **Ore:** Gold. **Ref: 13,** p. 174.

Murray Placer

Loc: Sec. 11, (30-28E), Kartar area. **Owner:** Gilbert V. Murray, Okanogan, Wash. (1951). **Ore:** Gold. **Prod:** Has produced intermittently. **Ref: 133,** p. 37. **158.**

Similkameen Placers (58)

Loc: Along the Similkameen R. between Oroville and Nighthawk. **Access:** Road from Nighthawk or Oroville. **Ore:** Gold. **Deposit:** Gold found as flake gold, shot gold, and nuggets in the river bars and lower terraces. **Prod:** Reportedly $500,000 in the few years following 1859. Intermittent to 1955. **Ref: 12,** pp. 6-7, 94. **13,** p. 173. **54,** p. 23. **154,** p. 96.

Similkameen Falls Placer (59)

Loc: At Similkameen Falls. **Ore:** Gold. **Ore min:** Free gold, ilmenite, magnetite, hematite. **Assays:** A cu. yd. of gravel yielded 110 lb. of conc. which showed 1,664 lb. magnetite, 160 lb. hematite, 152 lb. ilmenite, $31.40 gold per ton (1905). **Ref: 38-A,** p. 1218-1219. **126,** p. 15.

Strawberry Creek Placer

(See Crounse Placer.)

Walker Placer (104)

Loc: Secs. 13 and 14, (40-29E), Myers Cr. dist. **Ore:** Gold. **Prod:** 1930. **Ref: 97,** 1930, p. 673.

PACIFIC COUNTY

Fort Canby Placer (16)

Loc: Mouth of Columbia R., near Fort Canby. **Ore:** Gold. **Deposit:** Sand. **Assays:** 822 lb. magnetite, 240 lb. ilmenite, 81¢ gold per ton. **Ref: 38-A,** pp. 1218-1219. **126,** p. 15.

Ocean Park Placer (15)

Loc: Ocean beach at Ocean Park. **Ore:** Gold. **Deposit:** Sand. **Assays:** 22 lb. magnetite, 4 lb. ilmenite, 87¢ gold per ton. **Ref: 38-A,** pp. 1218-1219. **126,** p. 15.

Sand Island Placer (17)

Loc: Island at mouth of Columbia R., just S. of Ilwaco. **Ore:** Gold. **Deposit:** Sand. **Assays:** 160 lb. magnetite, 68 lb. ilmenite, 2 lb. zircon, $1.51 gold per ton. **Ref: 38-A,** pp. 1178, 1218-1219. **126,** p. 15.

PEND OREILLE COUNTY

Harvey Bar Placer (120)

Loc: On Pend Oreille R. in N. center sec. 26, (40-43E). **Access:** ½ mi. W. of Slate Cr. road. **Ore:** Gold. **Deposit:** Placer gravel on pockety dolomite bedrock. **Prod:** Intermittent. **Ref:** 128, p. 78.

O'Sullivan Creek Placer

(See Sullivan Creek Placer.)

Schierding Placer (119)

Loc: On E. side Pend Oreille R. just below mouth of Z Canyon gorge. **Elev:** 1,900 ft. **Access:** 2½ mi. by road W. of Crescent Lk. **Prop:** 120 acres of patented land. **Owner:** William Schierding and associates, Metaline Falls, Wash. (1943). **Ore:** Gold. **Deposit:** Well-rounded river gravel, generally less than 6 in. in dia., but some boulders up to 18 in. **Improv:** Dragline scraper, revolving grizzly, a screen and waterjet, and 2 sets of sluice boxes lined with riffles and blankets. Dismantled (1938). **Assays:** Small colors of flat gold with well-rounded edges. One nugget of 2 pennyweight. **Prod:** 1935. **Ref:** 29, p. 51. 128, p. 78.

Schultz Placer (122)

Loc: Near SE. cor. sec. 19, (39-43E), in draw near Schultz's cabin. **Ore:** Gold. **Deposit:** Surface gravels. **Assays:** Reportedly 50¢ to $1.50 per yd. **Prod:** Reportedly produced about 1900. **Ref:** 158.

Sullivan (O'Sullivan) Creek Placer (121)

Loc: On Sullivan Cr. E. of Metaline Falls. **Ore:** Gold. **Assays:** Nuggets valued at $20 each have been found. **Prod:** Reportedly several hundred thousand dollars. **Ref:** 12, pp. 8-9, 95. 13, p. 173.

PIERCE COUNTY

Ogren Placer (50)

Loc: Sec. 12, (17-10E), Summit dist. **Owner:** R. E. Ogren and Walter Zelepusa, Enumclaw, Wash. (1952 ——). **Ore:** Gold. **Ref:** 133, p. 38.

Silver Creek Placer (51)

Loc: NE¼ sec. 25, (17-10E), on headwaters of Silver Cr. **Elev:** 4,500 ft. **Access:** Road up Silver Cr. **Ore:** Gold. **Deposit:** Placer deposit from which considerable coarse gold has reputedly been recovered. **Improv:** Sluice boxes. **Assays:** Recoveries over a 10-yr. period av. $1.25 per yd. **Prod:** About 1920 to 1930. **Ref:** 158.

SKAGIT COUNTY

Anacortes Placer (29)

Loc: Near Anacortes. **Ore:** Gold. **Deposit:** Placer (probably beach). **Assays:** 2 samples showed 715 and 1,137 lb. chromite, 41¢ and 14¢ gold per ton. **Ref: 38-A,** pp. 1218-1219. **126.**

Day Creek Placer (30)

Loc: Day Cr. **Ore:** Gold. **Prod:** 1939. **Ref: 158.**

Ruby Creek Placer

Loc: Near headwaters of Ruby Cr. **Ore:** Gold. **Deposit:** First placer gold discovery west of the Cascade Mtns. **Assays:** Reportedly as high as $28 per yd. **Ref: 12,** pp. 7-8, 94. **13,** pp. 171-172. **145,** p. 95.

SKAMANIA COUNTY

Hudson and Meyers Placer (52)

Loc: On McCoy Cr., probably sec. 15, (10-8E). **Ore:** Gold. **Ref: 158.**

Texas Gulch Placer (54)

Loc: E½ sec. 25, (4-5E). **Ore:** Gold. **Ref: 158.**

SNOHOMISH COUNTY

Alpha and Beta Placers (35)

Loc: SW. cor. sec. 32, SE. cor. sec. 31, (30-10E), and E½ sec. 6, (29-10E). **Elev:** 2,000 ft. **Prop:** 2 patented claims. **Owner:** Sykes estate (1943). **Ore:** Gold. **Deposit:** Sand and gravel in flat area below steep slopes at head of Williamson Cr. **Dev:** No visible pits or other digging. **Ref: 23,** p. 50.

Aristo Placer (37)

Loc: Sec. 17, (28-8E), on Sultan R. just below the mouth of Sultan R. Canyon. **Owner:** H. E. Rathje, Sultan, Wash. **Ore:** Gold. **Prod:** Small amount produced by primitive methods and small hydraulic unit. **Ref: 14,** p. 50. **23,** p. 13.

Bench Placer (42)

Loc: Sec. 19, (28-11E), 800 ft. from junction of Silver Cr. with N. Fk. of Skykomish R. **Owner:** Mrs. Belle Fowler, Index, Wash. **Ore:** Gold. **Assays:** 80¢ to $1.00 per yd. **Ref: 158.**

Beta Placer

(See Alpha and Beta Placers.)

Darrington Placer (31)

Loc: NW¼ sec. 23, (32-9E), near town of Darrington on the W. side of Sauk R. **Ore:** Gold. **Ref: 14,** p. 50.

Deer Creek Placer (32)

Loc: On Deer Cr. near Darrington. **Ore:** Gold. **Ref:** 14, p. 50.

Gold Bar Placer (41)

Loc: Secs. 6, 7, and 8, (27-9E), on Skykomish R. **Ore:** Gold. **Prod:** Small amount. **Ref:** 14, p. 50.

Granite Falls Placer (33)

Loc: NW¼ sec. 8, (30-7E), on N. side of the S. Fk. of Stilaguamish R. about 1 mi. NE. of Granite falls. **Ore:** Gold. **Ref:** 14, pp. 50-51.

Horseshoe Bend Placer (38)

Loc: Sec. 8, (28-8E), on Sultan R. 5 mi. N. of the town of Sultan. **Prop:** 157 acres of patented ground. **Ore:** Coarse gold. **Dev:** Considerable amount of work. **Assays:** Av. values 25¢ to 40¢ per yd. **Prod:** Several thousand dollars. **Ref:** 14, p. 51. 63, p. 25.

McCloud Placer

(See Sultan Canyon Placer.)

Peterson Placer (34)

Loc: Near center sec. 9, (30-7E), on Stilaguamish R. **Prop:** Placer claim. **Owner:** Ed Peterson, East Stanwood, Wash. (1934). **Ore:** Gold. **Deposit:** Gold-bearing gravels being cyanided in 1934. **Ref:** 158.

Phoenix Placer (43)

Loc: NW¼ sec. 29, (28-11E), on Howard Cr. **Prop:** 1 claim. **Ore:** Gold. **Ref:** 14, p. 51.

Sailors Bar Placer

Loc: On Sultan R. **Ore:** Gold. **Prod:** $6,000. **Ref:** 14, p. 51. 63, p. 25.

Shirley Placer

Loc: On the N. Fk. of Skykomish R. near Galena. **Ore:** Gold. **Prod:** Small amount in 1934. **Ref:** 14, p. 51. 97, 1935, p. 354.

Stilaguamish River Placers

Loc: Along Stilaguamish R. **Ore:** Gold. **Assays:** Said to have been some rich placer deposits. **Ref:** 13, p. 172. 14, p. 51.

Sultan Placer (40)

Loc: NW¼ sec. 5, (27-8E), on Skykomish R., S. of the town of Sultan. **Ore:** Gold. **Prod:** $200 up to 1934, $1,408 during 1934. **Ref:** 14, p. 51.

Sultan Canyon (McCloud) Placer (36)

Loc: SE¼SE¼ sec. 32, (29-8E) and NE¼ sec. 5, (28-8E). **Access:** Road. **Owner:** Frank Morgan, Falls City, Wash. (1952). **Ore:** Gold. **Prod:** 1951. **Ref: 133,** p. 39.

Sultan River Placer (39)

Loc: On Sultan R. at W. edge of the town of Sultan, probably in sec. 31, (28-8E). **Ore:** Gold. **Improv:** Suction dredge operation. **Prod:** Some in 1947. **Ref: 158.**

Sultan River Placers

Loc: Along Sultan R. from its source to its mouth. **Ore:** Gold. **Assays:** Av. yield said to be 40¢ to 85¢ per yd. **Ref: 12,** p. 95. **13.** p. 172. **63,** pp. 25-26.

STEVENS COUNTY

Blue Bar Placer (137)

Loc: Sec. 20, (31-37E), Kettle Falls dist. **Ore:** Gold. **Prod:** 1934. **Ref: 97,** 1935, p. 354.

Bossburg Bar Placer (126)

Loc: Sec. 25, (38-37E), on W. bank of Columbia R. **Owner:** Federal Land Bank (1934). **Ore:** Gold. **Deposit:** Large bench or high bar which had 30,000 cu. yd. of gravel av. 25¢ per yd. **Ref: 158.**

Brod-Hurst Placer (128)

Loc: Marcus dist. **Ore:** Gold. **Prod:** 1934. **Ref: 97,** 1935, p. 354. **104,** 6/30/35, p. 22.

China Bend Placer (124)

Loc: Lots 3 and 4, sec. 7, (38-38E), on SE. bank of Columbia R. **Ore:** Gold. **Assays:** 8 samples showed 8.7¢ to 32.5¢ per cu. yd. **Ref: 158.**

Collins Placer (132)

Loc: Along Columbia R., 2½ mi. from Daisy. **Owner:** J. H. Collins, Colville, Wash. (1938). **Ore:** Gold. **Assays:** Av. 20¢ per yd. **Prod:** Owner grossed about $200 daily from 1,000 cu. yd. of gravel in 1938. **Ref: 97,** 1939, p. 491. **113,** 6/16/38.

Dead Man's Eddy Placer

(See Nigger Creek Bar Placer.)

Evans Placer (118)

Loc: Northport to international boundary. Camp on W. bank of the reservoir just N. of Northport. **Prop:** Placer lease on 5 mi. of shoreline on the Columbia R. reservoir. **Owner:** Ray Evans, Northport, Wash. (1941). **Ore:** Gold. **Improv:** 2 buildings, a shop, and a steam power suction dredge (1941). **Assays:** Said to range from 10¢ to $35.00 per yd. **Prod:** Presumably produced prior to 1941. **Ref: 30,** p. 103.

Gibson Bar Placer (140)

Loc: Secs. 10 and 15, (29-35E), at Gibson Bar, on E. side of Columbia R. **Owner:** Harvey R. Cline Co. (1940). **Ore:** Gold. **Deposit:** Gravel av. 4 ft. thick covering about 100 acres. **Assays:** 13 test holes av. 14.3¢ per cu. yd. **Prod:** 1938-1940; mined in early days by Chinese. **Ref: 97,** 1939, p. 491; 1940, p. 479; 1941, p. 475. **158.**

Holsten Placer (130)

Loc: Sec. 29, (35-37E), on E. bank of Columbia R., 6 mi. below Kettle Falls. **Owner:** Mr. Holsten (1934). **Ore:** Gold. **Assays:** Small production in 1934 on gravels running 26¢ to 30¢ per cu. yd. **Ref: 158.**

Hurst Placer

(See Brod-Hurst Placer.)

Marcus Placer (129)

Loc: At Marcus. **Ore:** Gold. **Deposit:** River sand. **Assays:** 1,096 lb. magnetite, 56 lb. ilmenite, $12.61 gold per ton from a 5-lb. conc. from 1 yd. of gravel. **Ref: 38-A,** pp. 1218-1219. **126,** p. 15.

Nigger Bar Placer (117)

Loc: Secs. 14 and 21, (40-41E), 2 mi. S. of Canadian boundary, on W. side of Columbia R. **Ore:** Gold. **Deposit:** Bar 300 ft. wide and more than 1 mi. long is covered by a 2-ft. overburden of sand and large boulders. **Assays:** A 125-ft. strip 1,000 ft. long had 2½ ft. of gravel av. 26¢ per cu. yd. **Ref: 158.**

Nigger Creek Bar (Dead Man's Eddy) Placer (115)

Loc: Sec. 28, (40-40E), at mouth of Nigger Cr., on W. bank of Columbia R. **Owner:** Roberts Leasing Co. (1933-1934). **Ore:** Gold. **Deposit:** Large low bar, with no large boulders. **Assays:** Av. 27.9¢ per cu. yd. **Prod:** 1933-1934. **Ref: 97,** 1934, p. 297; 1935, p. 355. **158.**

Ninemile Bar Placer (125)

Loc: Sec. 15, (38-38E), on E. bank of Columbia R. **Owner:** State (1934). **Ore:** Gold. **Deposit:** Large bar. **Assays:** 8 tests showed 2.6¢ to $1.18 per cu. yd. **Ref: 158.**

Northport Bar Placer (116)

Loc: Sec. 36, (40-39E), on W. bank of Columbia R., opposite Northport. **Ore:** Gold. **Assays:** 7 test pits showed an av. of 2 ft. of gravel which ran 20.1¢ per cu. yd. **Ref: 158.**

Reed and Roberts Placer (114)

Loc: Sec. 20, (40-40E), about 1 mi. W. of Columbia R. **Owner:** Indian Department leasing to Messrs. Reed and Roberts (1934). **Ore:** Gold. **Deposit:** High bench. **Prod:** $17,500 reported from less than 2 acres of gravel 3 ft. deep in 1934. **Ref: 158.**

Roberts Placer

(See Reed and Roberts Placer.)

Valbush Bar Placer (127)

Loc: Secs. 16 and 21, (37-38E), on E. side of Columbia R. between Marcus and Bossburg. **Owner:** State, leasing to Messrs. Billings, Nobel, Brod, Price, Raymond, and Foster (1934). **Ore:** Gold. **Assays:** 25,000 cu. yd. produced in 1934 yielded an av. of 21.4¢ per yd. Another operator on the same bar recovered $3,486.00 from 16,000 cu. yd. in 1934. **Prod:** 1934, 1939, 1940. **Ref: 97,** 1940, p. 479; 1941, p. 475. **158.**

WHATCOM COUNTY

Alice Mae Placer (23)

Loc: Secs. 11 and 12, (37-14E), on Ruby Cr., 1¼ mi. W. of Beebe Ranger Station. **Ore:** Gold. **Ref: 158.**

Combination Placer (20)

Loc: Sec. 2, (37-17E), Slate Cr. dist. **Prop:** 1 patented claim. **Owner:** Reuben Pierson, Seattle, Wash. (1952). **Ore:** Gold. **Ref: 158.**

Farrar Placer (25)

Loc: On Ruby Cr. between Lime and Darlington Creeks. Approx. in sec. 36, (38-16E). **Access:** Road. **Prop:** Several claims along a 2-mi. stretch of Ruby Cr. **Owner:** Mrs. John Farrar (1948). **Ore:** Gold. **Prod:** Several hundred dollars per year up to 1948. **Ref: 158.**

Jackass Placer (19)

Loc: Sec. 10, (37-14E), ½ mi. below junction of Cut Out Cr. and Granite Cr. **Prop:** 1 claim. **Owner:** Ralph E. Pride, Bellingham, Wash. (1952). **Ore:** Gold. **Ref: 158.**

Johnnie S. Placer (27)

Loc: Sec. 17, (37-16E), at mouth of Granite Cr., Slate Cr. dist. **Prop:** 1 claim. **Owner:** Morley Bouck, Rockport, Wash. (1952). **Ore:** Gold. **Ref: 158.**

Lazy Tar Heel Placer (21)

Loc: Secs. 10, 11, and 12, (37-14E), Ruby Cr. dist. **Owner:** Walt Woodrich, Sedro Woolley, Wash. (1939). **Ore:** Gold. **Prod:** 1939. **Ref: 158.**

Nip and Tuck (Tanya) Placer (22)

Loc: Sec. 11, (37-14E), Slate Cr. dist., about 5 mi. S. of Ruby Cr. **Prop:** 5 placer claims (1952). **Owner:** F. A. Weihe, Bellingham, Wash. (1952). R. E. Johnson, Winthrop, Wash. (1942). **Ore:** Gold. **Assays:** 50¢ per yd. **Ref: 158.**

Old Discovery Placer (28)

Loc: On Ruby Cr. about 2 mi. above its confluence with Granite Cr. Probably SE¼ sec. 9, (37-16E). **Ore:** Gold. **Deposit:** Placer gold is found in old stream channels well above present stream level. **Prod:** Several thousand dollars in "old days." **Ref: 63, p. 58. 158.**

Ruby Creek Placer (26)

Loc: Sec. 3, (37-17E), Slate Cr. dist. **Owner:** Granite Creek Mining Co. (1906-1909). **Ore:** Gold. **Prod:** 1906, 1908. **Ref: 63, p. 58. 97, 1906, p. 368; 1908, p. 582. 114, no. 5, 1909, p. 91.**

Scougale Placer (18)

Loc: Sec. 36, (38-13E), Slate Cr. dist. **Ore:** Gold. **Prod:** $950 taken out in 6 weeks in 1897. **Ref: 63. p. 58.**

Tanya Placer

(See Nip and Tuck Placer.)

Woodrich Placer (24)

Loc: Approx. in sec. 25, (37-14E), 2 mi. up Canyon Cr. from Beebe's cabin on Granite Cr. **Access:** About 5 mi. by trail from the Ruby Cr. road. **Ore:** Gold. **Deposit:** An old gravel bar, 300 ft. long and 5 to 20 ft. thick, from which at least 2 nuggets valued at $30 to $40 each were taken. **Ref: 158.**

WHITMAN COUNTY

Indian Bar Placer (153)

Loc: Sec. 15, (14-41E), near Penawawa. **Ore:** Gold. **Prod:** 1934. **Ref: 97, 1935, p. 356.**

YAKIMA COUNTY

Elizabeth Placer (47)

Loc: Secs. 4 and 5, (17-11E), Summit dist. **Owner:** Gold Links Mining Co. (1939). **Ore:** Gold. **Ref: 158.**

Gold Hill Placer (49)

Loc: Sec. 35, (17-11E), Morse Cr. area. **Ore:** Gold. **Prod:** 1933, 1934. **Ref: 97, 1934, p. 298; 1935, p. 356.**

Gold Links Placer (48)

Loc: Secs. 4 and 5, (17-11E), Summit dist. **Prop:** 3 placer claims, 20 lode claims. **Owner:** Gold Links Mining Co. (1939). **Ore:** Gold. **Ref: 158.**

Morse Creek Placer

Loc: Near head of Morse Cr., Summit dist. **Owner:** H. L. Tucker and George Gibbs, Yakima, Wash. (1880-1882). **Ore:** Gold. **Deposit:** An $80 nugget was found, and $1 nuggets were not uncommon. **Assays:** Early operators made "good wages." **Ref:** **63,** p. 44.

Surveyors Creek Placer (53)

Loc: Sec. 14, (8-12E), on Klickitat R. **Ore:** Gold. **Ref: 158.**

BIBLIOGRAPHY FOR OCCURRENCES

This bibliography is extracted from one which is more complete, compiled to be included in a forthcoming report on all metallic minerals of the state and to be published as Inventory of Washington Minerals, Part 2, Metallic Minerals: Division of Mines and Geology Bulletin 37. (Part 1 of Bulletin 37, dealing with nonmetallic minerals, was published in 1949.) Each reference, here, is assigned a number, but these numbers in the present report are not consecutive because only those references pertaining to gold have been taken from the long list, applicable to all metals, of Bulletin 37.

1.   Alaska and Northwest Mining Journal; continued as Pacific and Alaskan Review.

1-A. American Institute of Mining and Metallurgical Engineers, Transactions.

1-B. ————, Mining Technology.

3.   Arnold, Ralph, Gold placers of the coast of Washington: U. S. Geol. Survey Bull. 260, pp. 154-157, 1905.

7.   Bancroft, Howland, The ore deposits of northeastern Washington: U. S. Geol. Survey Bull. 550, 210 pp., 1914.

11-A. Bethel, H. L., Geology of the southeastern part of the Sultan quadrangle, King County, Washington: Unpublished Ph. D. thesis, Univ. Washington library, 1951.

12. Bethune, G. A., Mines and minerals of Washington: Washington State Geologist 1st Ann. Rept., 122 pp., 1890.

13. ————, Mines and minerals of Washington: Washington Geol. Survey 2d Ann. Rept., 183 pp., 1892.

14. Broughton, W. A., Inventory of mineral properties in Snohomish County, Washington: Washington Div. Geology Rept. Inv. 6, 64 pp., 1942.

22. Carithers, Ward, Directory of Washington mining operations: Washington Div. Mines and Mining Inf. Circ. 8, 36 pp., 1943.

23. ————, and Guard, A. K., Geology and ore deposits of the Sultan Basin, Snohomish County, Washington: Washington Div. Mines and Geology Bull. 36, 90 pp., 1945.

23-A. Chappell, W. M., Geology of the Wenatchee quadrangle, Washington: Unpublished Ph. D. thesis, Univ. Washington library, 1936.

24. Coats, R. R., The ore deposits of the Apex Gold mine, Money Creek, King County, Washington: Unpublished thesis, Univ. Washington library, 1932.

26. Collier, A. J., Gold-bearing river sands of northeastern Washington: U. S. Geol. Survey Bull. 315, pp. 56-70, 1907.

28. Colville Engineering Co., Ferry County power survey: Ferry County P. U. D. Rept., 140 pp., 1941.

29. ————, Report on minerals in Pend Oreille County: Pend Oreille County P. U. D. Rept., 75 pp., 1941.

30. ————, Report on minerals in Stevens County: Stevens County P. U. D. Rept., 137 pp., 1941.

31. Cooper, C. L., Mining and milling methods and costs at Knob Hill mine, Republic, Washington: U. S. Bur. Mines Inf. Circ. 7123, 29 pp., 1940.

33. The Copper Handbook; continued as Mines Handbook and later as Mines Register.

37. Culver, H. E., and Broughton, W. A., Tungsten resources of Washington: Washington Div. Geology Bull. 34, 89 pp., 1945.

38-A. Day, D. T., and Richards, R. H., Useful minerals in the black sands of the Pacific slope: U. S. Geol. Survey Mineral Resources, 1905, pp. 1175-1258, 1906.

39. Dobson, P. G., First Thought mine: Unpublished thesis, Univ. Washington library, 1917.

40. Dorisy, C. E., Index of mineral occurrences in the state of Washington: Washington State Planning Council, Research Pub. no. 3, 47 pp., 1937.

43. Engineering and Mining Journal (from 1922 to 1927 this was published as Engineering and Mining Journal-Press).

46. Gage, H. L., The zinc-lead mines of Washington: U. S. Dept. Interior, Bonneville Power Adm., Market Development Section, 235 pp., Dec. 1941.

47. Geological Society of America Bulletin.

49. Green, S. H., Directory of Washington mining operations, 1944: Washington Div. Mines and Mining Inf. Circ. 9, 36 pp., 1944.

51. ————, Directory of Washington mining operations, 1947: Washington Div. Mines and Geology Inf. Circ. 13, 59 pp., 1947.

52. ————, Directory of Washington mining operations, 1948: Washington Div. Mines and Geology Inf. Circ. 16, 51 pp., 1948.

54. Handy, F. M., An investigation of the mineral deposits of northern Okanogan County: State College of Washington Dept. of Geology Bull. 100, 27 pp., 1916.

57. Hill, T. B., and Melrose, J. W., Directory of Washington metallic mining properties: Washington Div. Mines and Mining Inf. Circ. 5, 72 pp., 1940.

58. ————, Directory of Washington metallic mining properties: Washington Div. Mines and Mining Inf. Circ. 7, 74 pp., 1941.

60. Hobbs, S. W., and Pecora, W. T., Nickel-gold deposit near Mount Vernon, Skagit County, Washington: U. S. Geol. Survey Bull. 931-D, pp. 57-78, 1941.

63. Hodges, L. K., Mining in the Pacific Northwest: Seattle Post-Intelligencer, 116 pp., 1897.

67. Huntting, M. T., Inventory of mineral properties in Chelan County, Washington: Washington Div. Geology Rept. Inv. 9, 63 pp., 1943.

68. ————, Directory of Washington mining operations, 1949: Washington Div. Mines and Geology Inf. Circ. 17, 62 pp., 1949.

69. ————, Directory of Washington mining operations, 1950: Washington Div. Mines and Geology Inf. Circ. 18, 67 pp., 1950.

73. Jenkins, O. P., Lead deposits of Pend Oreille and Stevens Counties, Washington: Washington Div. Geology Bull. 31, 153 pp., 1924.

75. Jones, E. L., Jr., Reconnaissance of the Conconully and Ruby mining districts, Washington: U. S. Geol. Survey Bull. 640, pp. 11-36, 1916.

76. Jonte, J. H., The relationship of selenium and gold in ores from the Republic district: Unpublished M. S. thesis, State College of Washington library, 1942.

77. Journal of Geology.

78. Julihn, C. E., and Moon, L. B., Summary of Bureau of Mines exploration projects on deposits of raw material resources for steel production: U. S. Bur. Mines Rept. Inv. 3801, 35 pp., 1945.

88. Landes, Henry, Thyng, W. S., Lyon, D. A., and Roberts, Milnor, The metalliferous resources of Washington, except iron: Washington Geol. Survey Ann. Rept. for 1901, pt. 2, 123 pp., 1902.

91. McIntyre, A. W., Copper deposits of Washington: Am. Min. Cong., 9th Ann. Sess., Rept. Proc., pp. 238-250, 1907.

97. Mineral Resources: U. S. Geol. Survey to 1924. U. S. Bur. Mines after 1924. Continued after 1931 as Minerals Yearbook.

98. Mines Handbook; continued as Mines Register.

99. Mines and Minerals. Seattle.

100. Mining. Spokane.

104. Mining Journal.

105. Mining and Scientific Press.

106. Mining Truth; continued as Northwest Mining. Formerly Northwest Mining Truth.

107. Mining World. Chicago. Continued as Mining and Engineering World.

108. Mining World. San Francisco.

111. Northern Pacific Railway Co., Tabulated summary of special geologic reconnaissance reports of Washington and northern Idaho: Land Dept., Geol. Div., 19 pp., 1941.

112. Northwest Mines Handbook.

113. Northwest Mining. Formerly Mining Truth, and formerly Northwest Mining Truth.

114. Northwest Mining Journal.

115. Northwest Mining and Metallurgy.

116. Northwest Mining News.

117. Northwest Mining Truth; continued as Mining Truth and later as Northwest Mining.

119. Pacific Mining Journal.

122. Pardee, J. T., Geology and mineral deposits of the Colville Indian Reservation, Washington: U. S. Geol. Survey Bull. 677, 186 pp., 1918.

126. ————, Platinum and black sand in Washington: U. S. Geol. Survey Bull. 805, pp. 1-15, 1929.

128. Park, C. F., Jr., and Cannon, R. S., Jr., Geology and ore deposits of the Metaline quadrangle, Washington: U. S. Geol. Survey Prof. Paper 202, 81 pp., 1943.

129. Patty, E. N., The metal mines of Washington: Washington Geol. Survey Bull. 23, 366 pp., 1921.

130. ————, and Glover, S. L., The mineral resources of Washington with statistics for 1919: Washington Geol. Survey Bull. 21, 155 pp., 1921.

131. Patty, E. N., and Kelly, S. F., A geological and geophysical study of the Chelan nickel deposit, near Winesap, Washington: Am. Inst. Min. Met. Eng. Tech. Pub. 1953, pp. 1-10, 1946.

132. Purdy, C. P., Jr., Antimony occurrences of Washington: Washington Div. Mines and Geology Bull. 39, 186 pp., 1951.

133. ————, Directory of Washington mining operations, 1952: Washington Div. Mines and Geology Inf. Circ. 20, 75 pp., 1952.

135-A. Reagan, A. B., Some notes on the Olympic Peninsula, Washington: Kansas Acad. Sci. Trans., vol. 22, pp. 131-238, 1909.

139. Schroeder, M. C., Geology of the Bead Lake district, Pend Oreille County, Washington: Washington Div. Mines and Geology Bull. 40, 57 pp., 1952.

141. Shedd, Solon, The mineral resources of Washington with statistics for 1922: Washington Div. Geology Bull. 30, pp. 1-183, 1924.

143. Smith, G. O., Gold mining in central Washington: U. S. Geol. Survey Bull. 213, pp. 76-80, 1903.

144. ————, U. S. Geol. Survey Geol. Atlas, Mount Stuart folio (no. 106), pp. 8-9, 1904.

145. ————, and Calkins, F. C., A geological reconnaissance across the Cascade range near the forty-ninth parallel: U. S. Geol. Survey Bull. 235, pp. 94-96, 1904.

146. ————, U. S. Geol. Survey Geol. Atlas, Snoqualmie folio (no. 139), pp. 13-14, 1906.

147. Smith, W. S., Petrology and economic geology of the Skykomish Basin, Washington: Colorado School of Mines Quarterly, vol. 36, pp. 154-188, Jan. 1915.

149. Spurr, J. E., The ore deposits of Monte Cristo, Washington: U. S. Geol. Survey 22d Ann. Rept., pt. 2, pp. 777-865, 1901.

150. Stebbins, R. H., Directory of Washington mining operations, 1951: Washington Div. Mines and Geology Inf. Circ. 19, 75 pp., 1951.

153. Umpleby, J. B., Geology and ore deposits of Republic mining district: Washington Geol. Survey Bull. 1, 65 pp., 1910.

154. ————, Geology and ore deposits of the Myers Creek mining district: Washington Geol. Survey Bull. 5, pt. 1, pp. 1-52, 1911.

157. U. S. Bureau of Mines data.

158. Washington Div. Mines and Geology unpublished data.

159. Washington, State of, Mines and Mining: Washington Bur. of Statistics, Agriculture, and Immigration Biennial Rept. 1903, pp. 124-138, 1903.

161. Weaver, C. E., Geology and ore deposits of the Blewett mining district: Washington Geol. Survey Bull. 6, 104 pp., 1911.

164. ————, The mineral resources of Stevens County: Washington Geol. Survey Bull. 20, 350 pp., 1920.

173. Woolf, J. A., and Towne, A. P., Refractory gold- and silver-bearing flotation concentrates from Knob Hill mine, Republic, Washington: U. S. Bur. Mines Rept. Inv. 3765, pp. 51-63, 1944.

174. Wright, L. B., Geologic relations and new ore bodies of the Republic district, Washington: Am. Inst. Min. Met. Eng. Trans., vol. 178, pp. 264-282, 1948.

175. Youngberg, E. A., and Wilson, T. L., The geology of the Holden mine: Econ. Geol., vol. 47, no. 1, pp. 1-12, 1952.

INDEX

A